BLACKWELL'S
UNDERGROUND CLINICAL VIGNETTES

BIOCHEMISTRY, 3E

BLACKWELL'S
UNDERGROUND CLINICAL VIGNETTES

BIOCHEMISTRY, 3E

VIKAS BHUSHAN, MD
University of California, San Francisco, Class of 1991
Series Editor, Diagnostic Radiologist

VISHAL PALL, MBBS
Government Medical College, Chandigarh, India, Class of 1996
Series Editor, U. of Texas, Galveston, Resident in Internal Medicine &
Preventive Medicine

TAO LE, MD
University of California, San Francisco, Class of 1996

JOSE M. FIERRO, MD
La Salle University, Mexico City

HOANG NGUYEN, MD, MBA
Northwestern University, Class of 2001

**Blackwell
Science**

CONTRIBUTORS

Tisha Wang
University of Texas Medical Branch, Class of 2002

Fadi Abu Shahin, MD
University of Damascus, Syria, Class of 1999

Vipal Soni, MD
UCLA School of Medicine, Class of 1999

© 2002 by Blackwell Science, Inc.

Editorial Offices:

Commerce Place, 350 Main Street, Malden,
    Massachusetts 02148, USA

Osney Mead, Oxford OX2 0EL, England

25 John Street, London WC1N 2BS, England

23 Ainslie Place, Edinburgh EH3 6AJ, Scotland

54 University Street, Carlton, Victoria 3053,
    Australia

Other Editorial Offices:

Blackwell Wissenschafts-Verlag GmbH,
    Kurfürstendamm 57, 10707 Berlin, Germany

Blackwell Science KK, MG Kodenmacho Building,
    7-10 Kodenmacho Nihombashi, Chuo-ku,
    Tokyo 104, Japan

Iowa State University Press, A Blackwell Science
    Company, 2121 S. State Avenue, Ames, Iowa
    50014-8300, USA

Distributors:

The Americas

Blackwell Publishing

c/o AIDC

P.O. Box 20

50 Winter Sport Lane

Williston, VT 05495-0020

(Telephone orders: 800-216-2522;
    fax orders: 802-864-7626)

Australia

Blackwell Science Pty, Ltd.

54 University Street

Carlton, Victoria 3053

(Telephone orders: 03-9347-0300;
    fax orders: 03-9349-3016)

Outside The Americas and Australia

Blackwell Science, Ltd.

c/o Marston Book Services, Ltd.

P.O. Box 269

Abingdon

Oxon OX14 4YN

England

(Telephone orders: 44-01235-465500;
    fax orders: 44-01235-465555)

Acquisitions: Laura DeYoung

Development: Amy Nuttbrock

Production: Lorna Hind and Shawn Girsberger

Manufacturing: Lisa Flanagan

Marketing Manager: Kathleen Mulcahy

Cover design by Leslie Haimes

Interior design by Shawn Girsberger

Typeset by TechBooks

Printed and bound by Capital City Press

**Blackwell's Underground Clinical Vignettes:
    Biochemistry, 3e**

ISBN 0-632-04545-0

Printed in the United States of America

02 03 04 05 5 4 3 2 1

The Blackwell Science logo is a trade mark of
Blackwell Science Ltd., registered at the United
Kingdom Trade Marks Registry

Library of Congress Cataloging-in-Publication Data

Bhushan, Vikas.

Blackwell's underground clinical vignettes.

Biochemistry / author, Vikas Bhushan. – 3rd ed.

    p. ; cm. – (Underground clinical vignettes)

Rev. ed. of: Biochemistry / Vikas Bhushan ... [et al.].

2nd ed. c1999. ISBN 0-632-04545-0 (pbk.)

1. Biochemistry – Case studies. 2. Physicians –
Licenses – United States – Examinations –
Study guides.

    [DNLM: 1. Biochemistry – Case Report.

2. Biochemistry – Problems and Exercises.

QU 18.2 B575b 2002] I. Title: Underground clinical
vignettes. Biochemistry. II. Title: Biochemistry.

III. Biochemistry. IV. Title. V. Series.

    RB112.5 .B48 2002

    572.8'076–dc21

2001004933

CONTENTS

ACKNOWLEDGMENTS

Throughout the production of this book, we have had the support of many friends and colleagues. Special thanks to our support team including Anu Gupta, Andrea Fellows, Anastasia Anderson, Srishti Gupta, Mona Pall, Jonathan Kirsch and Chirag Amin. For prior contributions we thank Gianni Le Nguyen, Tarun Mathur, Alex Grimm, Sonia Santos and Elizabeth Sanders.

We have enjoyed working with a world-class international publishing group at Blackwell Science, including Laura DeYoung, Amy Nuttbrock, Lisa Flanagan, Shawn Girsberger, Lorna Hind and Gordon Tibbitts. For help with securing images for the entire series we also thank Lee Martin, Kristopher Jones, Tina Panizzi and Peter Anderson at the University of Alabama, the Armed Forces Institute of Pathology, and many of our fellow Blackwell Science authors.

For submitting comments, corrections, editing, proofreading, and assistance across all of the vignette titles in all editions, we collectively thank:

Tara Adamovich, Carolyn Alexander, Kris Alden, Henry E. Aryan, Lynman Bacolor, Natalie Barteneva, Dean Bartholomew, Debashish Behera, Sumit Bhatia, Sanjay Bindra, Dave Brinton, Julianne Brown, Alexander Brownie, Tamara Callahan, David Canes, Bryan Casey, Aaron Caughey, Hebert Chen, Jonathan Cheng, Arnold Cheung, Arnold Chin, Simion Chiosea, Yoon Cho, Samuel Chung, Gretchen Conant, Vladimir Coric, Christopher Cosgrove, Ronald Cowan, Karekin R. Cunningham, A. Sean Dalley, Rama Dandamudi, Sunit Das, Ryan Armando Dave, John David, Emmanuel de la Cruz, Robert DeMello, Navneet Dhillon, Sharmila Dissanaike, David Donson, Adolf Etchegaray, Alea Eusebio, Priscilla A. Frase, David Frenz, Kristin Gaumer, Yohannes Gebreegziabher, Anil Gehi, Tony George, L.M. Gotanco, Parul Goyal, Alex Grimm, Rajeev Gupta, Ahmad Halim, Sue Hall, David Hasselbacher, Tamra Heimert, Michelle Higley, Dan Hoit, Eric Jackson, Tim Jackson, Sundar Jayaraman, Pei-Ni Jone, Aarchan Joshi, Rajni K. Jutla, Faiyaz Kapadi, Seth Karp, Aaron S. Kesselheim, Sana Khan, Andrew Pin-wei Ko, Francis Kong, Paul Konitzky, Warren S. Krackov, Benjamin H.S. Lau, Ann LaCasce, Connie Lee, Scott Lee, Guillermo Lehmann, Kevin Leung, Paul Levett, Warren Levinson, Eric Ley, Ken Lin,

Pavel Lobanov, J. Mark Maddox, Aram Mardian, Samir Mehta, Gil Melmed, Joe Messina, Robert Mosca, Michael Murphy, Vivek Nandkarni, Siva Naraynan, Carvell Nguyen, Linh Nguyen, Deanna Nobleza, Craig Nodurft, George Noumi, Darin T. Okuda, Adam L. Palance, Paul Pamphrus, Jinha Park, Sonny Patel, Ricardo Pietrobon, Riva L. Rahl, Aashita Randeria, Rachan Reddy, Beatriu Reig, Marilou Reyes, Jeremy Richmon, Tai Roe, Rick Roller, Rajiv Roy, Diego Ruiz, Anthony Russell, Sanjay Sahgal, Urmimala Sarkar, John Schilling, Isabell Schmitt, Daren Schuhmacher, Sonal Shah, Fadi Abu Shahin, Mae Sheikh-Ali, Edie Shen, Justin Smith, John Stulak, Lillian Su, Julie Sundaram, Rita Suri, Seth Sweetser, Antonio Talayero, Merita Tan, Mark Tanaka, Eric Taylor, Jess Thompson, Indi Trehan, Raymond Turner, Okafo Uchenna, Eric Uyguanco, Richa Varma, John Wages, Alan Wang, Eunice Wang, Andy Weiss, Amy Williams, Brian Yang, Hany Zaky, Ashraf Zaman and David Zipf.

For generously contributing images to the entire *Underground Clinical Vignette* Step 1 series, we collectively thank the staff at Blackwell Science in Oxford, Boston, and Berlin as well as:

- Axford, J. *Medicine.* Osney Mead: Blackwell Science Ltd, 1996. Figures 2.14, 2.15, 2.16, 2.27, 2.28, 2.31, 2.35, 2.36, 2.38, 2.43, 2.65a, 2.65b, 2.65c, 2.103b, 2.105b, 3.20b, 3.21, 8.27, 8.27b, 8.77b, 8.77c, 10.81b, 10.96a, 12.28a, 14.6, 14.16, 14.50.

- Bannister B, Begg N, Gillespie S. *Infectious Disease, 2ⁿᵈ Edition.* Osney Mead: Blackwell Science Ltd, 2000. Figures 2.8, 3.4, 5.28, 18.10, W5.32, W5.6.

- Berg D. *Advanced Clinical Skills and Physical Diagnosis.* Blackwell Science Ltd., 1999. Figures 7.10, 7.12, 7.13, 7.2, 7.3, 7.7, 7.8, 7.9, 8.1, 8.2, 8.4, 8.5, 9.2, 10.2, 11.3, 11.5, 12.6.

- Cuschieri A, Hennessy TPJ, Greenhalgh RM, Rowley DA, Grace PA. *Clinical Surgery.* Osney Mead: Blackwell Science Ltd, 1996. Figures 13.19, 18.22, 18.33.

- Gillespie SH, Bamford K. *Medical Microbiology and Infection at a Glance.* Osney Mead: Blackwell Science Ltd, 2000. Figures 20, 23.

- Ginsberg L. *Lecture Notes on Neurology, 7ᵗʰ Edition.* Osney Mead: Blackwell Science Ltd, 1999. Figures 12.3, 18.3, 18.3b.

- Elliott T, Hastings M, Desselberger U. *Lecture Notes on Medical Microbiology, 3ʳᵈ Edition.* Osney Mead: Blackwell Science Ltd, 1997. Figures 2, 5, 7, 8, 9, 11, 12, 14, 15, 16, 17, 19, 20, 25, 26, 27, 29, 30, 34, 35, 52.

- Mehta AB, Hoffbrand AV. *Haematology at a Glance.* Osney Mead: Blackwell Science Ltd, 2000. Figures 22.1, 22.2, 22.3.

Please let us know if your name has been missed or misspelled and we will be happy to make the update in the next edition.

PREFACE TO THE 3RD EDITION

We were very pleased with the overwhelmingly positive student feedback for the 2nd edition of our *Underground Clinical Vignettes* series. Well over 100,000 copies of the UCV books are in print and have been used by students all over the world.

Over the last two years we have accumulated and incorporated **over a thousand "updates"** and improvements suggested by you, our readers, including:

- many additions of specific boards and wards testable content

- deletions of redundant and overlapping cases

- reordering and reorganization of all cases in both series

- a new master index by case name in each Atlas

- correction of a few factual errors

- diagnosis and treatment updates

- addition of 5–20 new cases in every book

- and the addition of clinical exam photographs within *UCV—Anatomy*

And most important of all, the third edition sets now include two brand new **COLOR ATLAS** supplements, one for each Clinical Vignette series.

- The *UCV–Basic Science Color Atlas* (*Step 1*) includes over 250 color plates, divided into gross pathology, microscopic pathology (histology), hematology, and microbiology (smears).

- The *UCV–Clinical Science Color Atlas* (*Step 2*) has over 125 color plates, including patient images, dermatology, and funduscopy.

Each atlas image is descriptively captioned and linked to its corresponding Step 1 case, Step 2 case, and/or Step 2 MiniCase.

How Atlas Links Work:

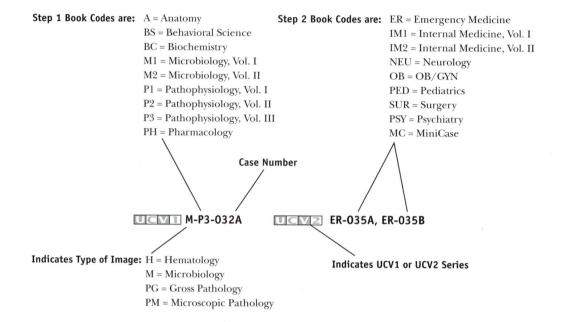

Step 1 Book Codes are:
A = Anatomy
BS = Behavioral Science
BC = Biochemistry
M1 = Microbiology, Vol. I
M2 = Microbiology, Vol. II
P1 = Pathophysiology, Vol. I
P2 = Pathophysiology, Vol. II
P3 = Pathophysiology, Vol. III
PH = Pharmacology

Step 2 Book Codes are:
ER = Emergency Medicine
IM1 = Internal Medicine, Vol. I
IM2 = Internal Medicine, Vol. II
NEU = Neurology
OB = OB/GYN
PED = Pediatrics
SUR = Surgery
PSY = Psychiatry
MC = MiniCase

Case Number

UCV1 M-P3-032A   UCV2 ER-035A, ER-035B

Indicates Type of Image:
H = Hematology
M = Microbiology
PG = Gross Pathology
PM = Microscopic Pathology

Indicates UCV1 or UCV2 Series

- If the Case number (032, 035, etc.) is not followed by a letter, then there is only one image. Otherwise A, B, C, D indicate up to 4 images.

Bold Faced Links: In order to give you access to the largest number of images possible, we have chosen to cross link the Step 1 and 2 series.

- If the link is bold-faced this indicates that the link is direct (i.e., Step 1 Case with the Basic Science Step 1 Atlas link).

- If the link is not bold-faced this indicates that the link is indirect (Step 1 case with Clinical Science Step 2 Atlas link or vice versa).

We have also implemented a few structural changes upon your request:

- Each current and future edition of our popular *First Aid for the USMLE Step 1* (Appleton & Lange/McGraw-Hill) and *First Aid for the USMLE Step 2* (Appleton & Lange/McGraw-Hill) book will be linked to the corresponding UCV case.

- We eliminated UCV → First Aid links as they frequently become out of date, as the *First Aid* books are revised yearly.

- The Color Atlas is also specially designed for quizzing—captions are descriptive and do not give away the case name directly.

We hope the updated UCV series will remain a unique and well-integrated study tool that provides compact clinical correlations to basic science information. They are designed to be easy and fun (comparatively) to read, and helpful for both licensing exams and the wards.

We invite your corrections and suggestions for the fourth edition of these books. For the first submission of each factual correction or new vignette that is selected for inclusion in the fourth edition, you will receive a personal acknowledgment in the revised book. If you submit over 20 high-quality corrections, additions or new vignettes we will also consider **inviting you to become a "Contributor" on the book of your choice**. If you are interested in becoming a potential "Contributor" or "Author" on a future UCV book, or working with our team in developing additional books, please also e-mail us your CV/resume.

We prefer that you submit corrections or suggestions via electronic mail to **UCVteam@yahoo.com**. Please include "Underground Vignettes" as the subject of your message. If you do not have access to e-mail, use the following mailing address: Blackwell Publishing, Attn: UCV Editors, 350 Main Street, Malden, MA 02148, USA.

Vikas Bhushan
Vishal Pall
Tao Le
October 2001

HOW TO USE THIS BOOK

This series was originally developed to address the increasing number of clinical vignette questions on medical examinations, including the USMLE Step 1 and Step 2. It is also designed to supplement and complement the popular *First Aid for the USMLE Step 1* (Appleton & Lange/McGraw Hill) and *First Aid for the USMLE Step 2* (Appleton & Lange/McGraw Hill).

Each UCV 1 book uses a series of approximately 100 **"supra-prototypical" cases as a way to condense testable facts and associations**. The clinical vignettes in this series are designed to incorporate as many testable facts as possible into a cohesive and memorable clinical picture. The vignettes represent composites drawn from general and specialty textbooks, reference books, thousands of USMLE style questions and the personal experience of the authors and reviewers.

Although each case tends to present all the signs, symptoms, and diagnostic findings for a particular illness, **patients generally will not present with such a "complete" picture either clinically or on a medical examination**. Cases are not meant to simulate a potential real patient or an exam vignette. All the **boldfaced "buzzwords" are for learning purposes** and are not necessarily expected to be found in any one patient with the disease.

Definitions of selected important terms are placed within the vignettes in (SMALL CAPS) in parentheses. Other parenthetical remarks often refer to the pathophysiology or mechanism of disease. The format should also help students learn to present cases succinctly during oral "bullet" presentations on clinical rotations. The cases are meant to serve as a condensed review, not as a primary reference. The information provided in this book has been prepared with a great deal of thought and careful research. This book should not, however, be considered as your sole source of information. Corrections, suggestions and submissions of new cases are encouraged and will be acknowledged and incorporated when appropriate in future editions.

ABBREVIATIONS

5-ASA	5-aminosalicylic acid
ABGs	arterial blood gases
ABVD	adriamycin/bleomycin/vincristine/dacarbazine
ACE	angiotensin-converting enzyme
ACTH	adrenocorticotropic hormone
ADH	antidiuretic hormone
AFP	alpha fetal protein
AI	aortic insufficiency
AIDS	acquired immunodeficiency syndrome
ALL	acute lymphocytic leukemia
ALT	alanine transaminase
AML	acute myelogenous leukemia
ANA	antinuclear antibody
ARDS	adult respiratory distress syndrome
ASD	atrial septal defect
ASO	anti-streptolysin O
AST	aspartate transaminase
AV	arteriovenous
BE	barium enema
BP	blood pressure
BUN	blood urea notrogen
CAD	coronary artery disease
CALLA	common acute lymphoblastic leukemia antigen
CBC	complete blood count
CHF	congestive heart failure
CK	creatine kinase
CLL	chronic lymphocytic leukemia
CML	chronic myelogenous leukemia
CMV	cytomegalovirus
CNS	central nervous system
COPD	chronic obstructive pulmonary disease
CPK	creatine phosphokinase
CSF	cerebrospinal fluid
CT	computed tomography
CVA	cerebrovascular accident
CXR	chest x-ray
DIC	disseminated intravascular coagulation
DIP	distal interphalangeal
DKA	diabetic ketoacidosis
DM	diabetes mellitus
DTRs	deep tendon reflexes
DVT	deep venous thrombosis

EBV	Epstein–Barr virus
ECG	electrocardiography
Echo	echocardiography
EF	ejection fraction
EGD	esophagogastroduodenoscopy
EMG	electromyography
ERCP	endoscopic retrograde cholangiopancreatography
ESR	erythrocyte sedimentation rate
FEV	forced expiratory volume
FNA	fine needle aspiration
FTA-ABS	fluorescent treponemal antibody absorption
FVC	forced vital capacity
GFR	glomerular filtration rate
GH	growth hormone
GI	gastrointestinal
GM-CSF	granulocyte macrophage colony stimulating factor
GU	genitourinary
HAV	hepatitis A virus
hcG	human chorionic gonadotrophin
HEENT	head, eyes, ears, nose, and throat
HIV	human immunodeficiency virus
HLA	human leukocyte antigen
HPI	history of present illness
HR	heart rate
HRIG	human rabies immune globulin
HS	hereditary spherocytosis
ID/CC	identification and chief complaint
IDDM	insulin-dependent diabetes mellitus
Ig	immunoglobulin
IGF	insulin-like growth factor
IM	intramuscular
JVP	jugular venous pressure
KUB	kidneys/ureter/bladder
LDH	lactate dehydrogenase
LES	lower esophageal sphincter
LFTs	liver function tests
LP	lumbar puncture
LV	left ventricular
LVH	left ventricular hypertrophy
Lytes	electrolytes
MCHC	mean corpuscular hemoglobin concentration
MCV	mean corpuscular volume
MEN	multiple endocrine neoplasia

MGUS	monoclonal gammopathy of undetermined significance
MHC	major histocompatibility complex
MI	myocardial infarction
MOPP	mechlorethamine/vincristine (Oncovorin)/procarbazine/prednisone
MR	magnetic resonance (imaging)
NHL	non-Hodgkin's lymphoma
NIDDM	non-insulin-dependent diabetes mellitus
NPO	nil per os (nothing by mouth)
NSAID	nonsteroidal anti-inflammatory drug
PA	posteroanterior
PIP	proximal interphalangeal
PBS	peripheral blood smear
PE	physical exam
PFTs	pulmonary function tests
PMI	point of maximal intensity
PMN	polymorphonuclear leukocyte
PT	prothrombin time
PTCA	percutaneous transluminal angioplasty
PTH	parathyroid hormone
PTT	partial thromboplastin time
PUD	peptic ulcer disease
RBC	red blood cell
RPR	rapid plasma reagin
RR	respiratory rate
RS	Reed–Sternberg (cell)
RV	right ventricular
RVH	right ventricular hypertrophy
SBFT	small bowel follow-through
SIADH	syndrome of inappropriate secretion of ADH
SLE	systemic lupus erythematosus
STD	sexually transmitted disease
TFTs	thyroid function tests
tPA	tissue plasminogen activator
TSH	thyroid-stimulating hormone
TIBC	total iron-binding capacity
TIPS	transjugular intrahepatic portosystemic shunt
TPO	thyroid peroxidase
TSH	thyroid-stimulating hormone
TTP	thrombotic thrombocytopenic purpura
UA	urinalysis
UGI	upper GI
US	ultrasound

VDRL	Venereal Disease Research Laboratory
VS	vital signs
VT	ventricular tachycardia
WBC	white blood cell
WPW	Wolff–Parkinson–White (syndrome)
XR	x-ray

ID/CC A 16-year-old girl is referred to an endocrinologist owing to **lack of menses** (PRIMARY AMENORRHEA) and absence of pubic hair, axillary hair, and breast development (LACK OF SECONDARY SEXUAL CHARACTERISTICS).

HPI She also complains of frequent **headaches and ringing in her ears** (due to hypertension).

PE VS: **hypertension** (BP 160/105). PE: funduscopic exam normal; no lymphadenopathy; no hepatosplenomegaly; absence of breast tissue; no abdominal or pelvic masses palpable; no axillary or pubic hair; vulvar labia normal.

Labs Lytes: **hypokalemia**; hypernatremia. ABGs: metabolic **alkalosis** (due to mineralocorticoid action of 11-deoxycorticosterone and corticosterone). Suppressed renin; increase in urinary gonadotropins (due to attempt to compensate for lack of sex hormones); diminished 17-ketosteroids (product of sex hormones); increased progesterone, pregnenolone, 11-deoxycorticosterone, and corticosterone.

Treatment Glucocorticoids. Sex hormones.

Discussion A decrease in 17-α-hydroxylase produces an increase in 11-deoxycorticosterone and corticosterone (due to shifting metabolism of sex hormones to aldosterone pathway); renin is suppressed (due to aldosterone negative feedback). Females fail to develop secondary sexual characteristics; males develop ambiguous external genitalia (MALE PSEUDOHERMAPHRODITISM).

1 **17-ALPHA-HYDROXYLASE DEFICIENCY**

ID/CC A neonatal boy is brought to the pediatrician by his father, who recently discovered that his son **does not urinate through his penis**.

HPI The patient's father also reports that he cannot find his son's testes (due to cryptorchidism).

PE Penis small for age (MICROPHALLIA); testes located in inguinal canal bilaterally (CRYPTORCHIDISM); urinary meatus lies in perineum (HYPOSPADIAS); scrotal sac bifid.

Labs **Markedly reduced dihydrotestosterone with normal testosterone level.** Decreased 5-α-reductase activity.

Imaging Pelvic US: confirms cryptorchidism and absent uterus.

Treatment Psychosocial support. Gender assignment. Consider appropriate hormonal replacement therapy.

Discussion An **autosomal-recessive** disorder of virilization affecting genetic males. 5-α-reductase converts testosterone to dihydrotestosterone. Lack of type 2 isozyme of 5-alpha-reductase produces a decrease in dihydrotestosterone, which is responsible for virilization of the external genitalia.

ID/CC A 43-year-old white male comes to the emergency room complaining of severe retro-orbital **headache** (behind his eyes) along with **blurred vision**.

HPI He also complains of weakness over the past few months and an **increase in hat size** as well as an inability to wear his wedding ring (due to growth in finger width). His family also notes a **coarsening of his facial features and deepening of his voice**.

PE VS: hypertension (BP 150/100). PE: skin thick and oily; **prominent forehead and jaw**; enlarged tongue and widening gaps between teeth; **large hands and feet; bitemporal hemianopsia; cardiomegaly**; hepatosplenomegaly.

Labs **Hyperglycemia**; hyperphosphatemia; **increased IGF-1; increased levels of GH** that **fail to suppress** after oral glucose load. UA: increased urinary calcium.

Imaging XR: thickening of skull; erosion and **enlargement of sella turcica**; widening distal phalanges in hands and feet. MR, head: enlarged pituitary gland containing a 2 cm mass.

Gross Pathology Acidophilic adenoma of pituitary gland with ill-defined capsule exerts mass effects on pituitary and nearby optic chiasm; tumor rarely malignant.

Micro Pathology Densely packed, mature cells that are highly granulocytic and eosinophilic; stains highly for GH.

Treatment Transsphenoidal microsurgical adenomectomy is the treatment of choice; radiotherapy to reduce further growth of tumor. Medical therapy with octreotide and/or bromocriptine if surgery fails or is contraindicated.

Discussion The most common cause of acromegaly is **pituitary adenoma**. If excess GH secretion is present in **childhood, gigantism** appears; **in adults, acromegaly** appears. Headache and joint pain are early complaints; blurred vision and visual-field changes occur later. Almost every organ in the body increases in size; 25% of patients exhibit glucose intolerance. Visual field changes (e.g., bitemporal hemianopsia) may occur secondary to compression of the nerves of the optic chiasm by the tumor.

Atlas Links UCV1 PM-BC-003   UCV2 MC-084

ACROMEGALY

ID/CC A 37-year-old female is admitted to the internal medicine ward for evaluation of **increasing weakness** and intermittent episodes of **dizziness, nausea, and vomiting related to stress and exercise**.

HPI She is a vegetarian, takes no drugs or medications, and does not drink alcohol or smoke cigarettes. She reports an excessive **craving for salty foods** such as chips and salted peanuts.

PE VS: **tachycardia** (HR 110); **hypotension** (BP 90/65). PE: thin with dry mucous membranes; **pigmentation of buccal mucosa and palms of hands**; no neck masses; chest auscultation normal; no abdominal masses; no hepatosplenomegaly; no lymphadenopathy.

Labs CBC: normal. Lytes: **hyponatremia; hyperkalemia. Glucose low; increased BUN with normal creatinine**; amylase and LFTs normal; **high ACTH; low cortisol**.

Treatment Glucocorticoid and mineralocorticoid hormones. Hydrocortisone on an emergent basis.

Discussion Primary hypoadrenalism (ADDISON'S DISEASE) may be caused by **autoimmune mechanisms**, tuberculous infection, or sudden discontinuation of chronic steroid administration. Secondary hypoadrenalism is due to abnormalities of hypothalamic-pituitary function.

ADDISON'S DISEASE

ID/CC A **15-year-old** female is admitted to the hospital for evaluation of persistent **weakness** for the last 6 months that has been unresponsive to multivitamin treatment.

HPI She denies allergies, surgeries, psychological problems, transfusions, drug use, or any relevant past medical history.

PE VS: heart rate normal; no fever; **normal BP** (excludes primary hyperaldosteronism). PE: well hydrated; pupils equal and reactive to light and accommodation; no neck masses; no lymphadenopathy; chest normal; abdomen soft and nontender; no masses; neurologic exam normal; **no peripheral edema**; sexual development appropriate for age.

Labs CBC: normal. Lytes: **hyponatremia; hypokalemia**. ABGs: **metabolic alkalosis. Increased plasma renin** (excludes primary hyperaldosteronism); increased urinary excretion of prostaglandins.

Micro Pathology **Juxtaglomerular cell hyperplasia** on renal biopsy.

Treatment Indomethacin to decrease prostaglandin synthesis. Potassium chloride supplements and potassium-sparing diuretics.

Discussion Bartter's syndrome is a hereditary disorder characterized by a defective Na^+-K^+-$2Cl^-$ cotransporter in the thick ascending loop of Henle. This results in an impaired reabsorptive conservation mechanism of sodium (due to end-organ resistance to angiotensin), with urinary sodium wasting and a consequent increase in renin production (through increased renal prostaglandins). Therefore, there is an increase in aldosterone activity with hypokalemic alkalosis. Hypokalemia perpetuates the cycle by stimulating renin activity.

ID/CC   A newborn is evaluated by a neonatologist because the intern who performed the delivery **cannot tell whether the child is male or female** (AMBIGUOUS GENITALIA).

HPI   The child is also **lethargic** and lacks sufficient strength to suck on mother's milk adequately (due to salt wasting).

PE   Ambiguous external genitalia; **increase in size of clitoris; fusion of labia** to the point of resembling scrotal sac.

Labs   Lytes: hyponatremia; **hyperkalemia. Increase in 17-α-OH progesterone** and its metabolite, **pregnanetriol; increase in urinary 17-ketosteroids** (defect is distal to 17, 20-desmolase); elevated serum ACTH. Prenatal diagnosis is possible at 14 to 16 weeks (due to increase in 17-α-OH progesterone). Karyotype: 46,XX female.

Treatment   Cortisol, dehydrocorticosterone acetate if salt wasting is present.

Discussion   Lack of 21-hydroxylase causes a decrease in cortisol with a consequent increase in ACTH, which in turn produces hyperplasia of the adrenals—resulting in an increase in androgen production that gives rise to signs of female pseudohermaphroditism (as in this case) or enlarged genitalia in the male. May occur with or without salt wasting.

CONGENITAL ADRENAL HYPERPLASIA

ID/CC A 27-year-old white male complains of **excessive thirst** (POLYDIP-SIA) and **increased urination** (POLYURIA) with very diluted urine.

HPI The patient drinks several liters of water every day. He was well until this time. The patient also admits to frequent urination (including nocturia) of large volumes that are clear and watery.

PE VS: slight tachycardia. PE: mild dryness of mucous membranes; visual field testing normal; no papilledema; pupils equal and reactive.

Labs **Low urine specific gravity** (< 1.006); **low urine osmolarity** (< 200 mOsm/L); **elevated serum osmolality** (> 290 mOsm/L); **hypernatremia; water deprivation test** demonstrates inability to concentrate urine with fluid restriction (urinary osmolality continues to be low).

Imaging CT: may show masses or lesions in hypothalamus.

Treatment Central (primary) diabetes insipidus: intranasal desmopressin, diuretics and ADH-releasing drugs such as chlorpropamide, carbamazepine and clofibrate. Nephrogenic (secondary) diabetes insipidus: add indomethacin, amiloride, and/or hydrochlorothiazide.

Discussion Diabetes insipidus is caused by an ADH deficiency (PRIMARY) or by renal unresponsiveness to ADH (NEPHROGENIC OR SECONDARY). Primary diabetes insipidus can be caused by surgical, traumatic, or anoxic damage to the hypothalamus or pituitary stalk during pregnancy (SHEEHAN'S SYNDROME). Deficiency of ADH results in **renal loss of free water and hypernatremia**.

7 **DIABETES INSIPIDUS**

ID/CC A 28-year-old seamstress is admitted to the internal medicine ward because of malaise, **confusion, abdominal pain**, vomiting, and diarrhea.

HPI She is a known **insulin-dependent diabetic** (IDDM type I, juvenile onset). One day before her admission, she went out to celebrate her birthday and drank **alcohol** until she became intoxicated (she also forgot to administer insulin).

PE VS: tachycardia (HR 92); hypotension (BP 90/50) (due to hypovolemia); **rapid, deep breathing** (KUSSMAUL RESPIRATION). PE: **dehydration**; peripheral cyanosis; cold, **dry skin; peculiar fruity breath smell** (due to ketone bodies, acetoacetate, and β-OH-butyrate).

Labs CBC: leukocytosis (14,000) (without infection). Lytes: hyponatremia (130 mEq/L). ABGs: **markedly reduced bicarbonate** (10 mEq/L); **acidosis** (pH = 7.1). Increased ketones in blood; increased creatinine; **hyperglycemia; increased anion gap** (between 10 and 18) (anion gap is calculated as follows: $[Na + K]-[Cl + HCO_3]$); increased amylase (without pancreatitis). UA: glycosuria; ketonuria.

Treatment Nasogastric tube, correction of fluid deficit (caution owing to risk of producing cerebral edema). **Potassium**. Gradual lowering of glucose with **insulin**.

Discussion Ketoacidosis might be the first manifestation of diabetes. It is more common in insulin-dependent diabetics than hyperosmolar coma. It usually follows a period of physical or mental stress (e.g., MI, acute grief) or infection.

ID/CC A 58-year-old white **female** comes to see her internist because of the development of **polyuria, polydipsia** (due to hyperglycemia), and a **skin eruption** that comes and goes in different parts of her body (NECROLYTIC MIGRATORY RASH).

HPI She also complains of increasing intermittent **diarrhea, nausea, vomiting, weight loss**, and occasional weakness and dizziness.

PE VS: normal. PE: patient well hydrated; marked pallor; **erythematous rash on anterior chest, legs, and arms**; no neck masses; lungs clear to auscultation; heart sounds rhythmic; abdomen soft; no masses; no peritoneal signs; no lymphadenopathy.

Labs CBC: anemia (Hb 7.4 mg/dL). **Markedly increased serum glucagon** levels; **hyperglycemia**.

Imaging MR/CT: 2.5-cm enhancing mass in body and tail of pancreas; several liver metastases.

Treatment Surgical removal. Streptozocin if metastatic; insulin; prophylactic heparin and zinc (for skin rash); octreotide.

Discussion Glucagonoma is a pancreatic islet cell neoplasm (of α cells) that secretes abnormally high amounts of glucagon with resulting **symptomatic hyperglycemia**; it may also secrete gastrin, ACTH, and serotonin. Glucagonomas arise from α_2 islet cells in the pancreas, and the majority (> 70%) are malignant. Glucagonomas may also be associated with multiple endocrine neoplasia (MEN) type I.

Atlas Links UCV1 PM-BC-009A, PM-BC-009B

ID/CC A 24-year-old woman is referred to the endocrinologist because of concern over **excessive facial hair** along with hair on her central chest and thighs.

HPI The patient's **menses are regular**, with an average flow lasting 3 to 4 days. She is not taking any drugs.

PE Hirsutism noted; **no clitoromegaly** present (no evidence of virilization); no abdominal or pelvic mass palpable per abdomen or per vagina.

Labs **Normal total testosterone** levels; **normal DHEAS**; normal urine for 17-ketosteroids.

Imaging US, abdomen and pelvis: both adrenals and ovaries normal.

Treatment No therapy required.

Discussion Hirsutism that is disproportionate to the patient's ethnic background and is accompanied by normal periods is termed idiopathic. If testosterone and DHEAS levels are normal, the patient can be reassured that the condition is benign. If the onset of hirsutism is pubertal with irregular periods, the possibility of polycystic ovarian syndrome exists. Recent-onset hirsutism in an adult female, especially when associated with amenorrhea, requires complete investigation to exclude an adrenal or ovarian tumor.

ID/CC A 39-year-old woman is referred to an internist by her family practitioner because of persistent **hypertension** that has been **unresponsive to conventional treatment**; she also complains of profound **muscle weakness** (due to hypokalemia).

HPI She is a two-pack-a-day smoker who drinks occasionally.

PE VS: normal heart rate; **hypertension** (BP 200/100). PE: no pallor; **retinal hemorrhages, exudates, and AV nicking**; lungs clear; no heart murmurs; abdomen soft; no palpable masses; no lymphadenopathy.

Labs CBC: increased hematocrit. Lytes: **hypokalemia; hypernatremia** (secondary to hyperaldosteronism). ABGs: **high serum bicarbonate**. Glucose normal (vs. ectopic ACTH production). ECG: left ventricular hypertrophy and strain. UA: no proteinuria. **Aldosterone levels high; renin levels low** (primary hyperaldosteronism).

Imaging CT/MR: 1.7-cm enhancing left adrenal mass. NP-59 iodocholesterol scintigraphy positive. Adrenal venous sampling shows >10:1 ratio of aldosterone on the left versus right.

Micro Pathology Glomerulosa-cell benign **adrenal adenoma**.

Treatment Left adrenalectomy.

Discussion Primary hyperaldosteronism typically presents with hypertension, hypokalemia, hypernatremia, and increased bicarbonate due to increased secretion of aldosterone by an adrenal adenoma (CONN'S SYNDROME) or hyperplasia. Hypertension is characteristically **unresponsive to ACE inhibitors**. Surgically correctable causes of hypertension include Conn's syndrome, pheochromocytoma, renal artery stenosis, and coarctation of the aorta.

HYPERALDOSTERONISM—PRIMARY

ID/CC A 55-year-old menopausal female comes to see her internist because of progressive **constipation** and **excessive urination** over the past 2 months; she also complains of **palpitations** both at rest and during exercise.

HPI She has read "all about" osteoporosis during menopause and is afraid of developing it, so she has been taking **abundant calcium supplements** and vitamin D injections.

PE VS: heart rate 80 with skipped beats heard (VENTRICULAR EXTRASYSTOLES). PE: lungs clear; no neck masses; thyroid not palpable; no lymphadenopathy; **muscle weakness with hyporeflexia**; abdomen soft with decreased bowel sounds; no masses; no abnormal pigmentation; **soft tissue calcification** in skin of arms and legs.

Labs **Markedly increased serum calcium** (12 mg/dL) (always correct calcium level for serum albumin). Phosphorus normal (makes primary hyperparathyroidism less likely). ABGs: metabolic alkalosis. Increased BUN. ECG: **short Q-T**. No PTH-related protein detected.

Treatment Aggressive rehydration with normal saline; diuresis with furosemide to increase sodium and concomitant calcium excretion. Calcitonin, plicamycin, or bisphosphonates (etidronate, pamidronate) may be necessary.

Discussion Hypercalcemia may occur in hyperparathyroidism, milk-alkali syndrome, multiple myeloma, Addison's disease, sarcoidosis, prolonged immobilization, metastatic neoplastic disease (due to increased osteoclastic resorption), and primary neoplastic disease (due to production of a PTH-like substance). Fifty percent of serum calcium is bound to albumin; most of the rest is actively reabsorbed in the proximal tubule together with sodium. This reabsorption is decreased with expansion of extracellular fluid volume.

ID/CC A 53-year-old white **female** goes to her family doctor for a routine physical and is found to be **hypercalcemic**.

HPI She is asymptomatic except for mild **polyuria**.

PE VS: mild **hypertension**. PE: no neck masses; thyroid not palpable; no lymphadenopathy; lungs clear; heart sounds normal; abdomen soft; no masses.

Labs **Increased serum calcium; phosphorus low; elevated PTH; increased alkaline phosphatase**. UA: increased urinary calcium; elevated urinary cAMP and hydroxyproline levels. ABGs: **hyperchloremic metabolic acidosis** (normal anion gap). ECG: **short Q-T**.

Imaging XR: subperiosteal bone resorption; **cystic long-bone lesions** (BROWN TUMORS). Nuc: increased bone uptake on bone scan.

Gross Pathology Soft, round, well-encapsulated, yellowish-brown single parathyroid adenoma weighing 2 g.

Micro Pathology Chief cells within adenoma.

Treatment Surgical removal.

Discussion Primary hypersecretion of parathyroid hormone may be caused by an **adenoma** (vast majority of cases), chief-cell hyperplasia, or carcinoma of the parathyroid glands; it is commonly asymptomatic and is frequently recognized during routine physical exams. When it is symptomatic, peptic ulcer pain, polyuria, polydipsia, constipation, and pancreatitis may be the presenting symptoms. May be associated with multiple endocrine neoplasia (MEN) syndromes I and II.

Atlas Links ⬛UCV1⬛ PM-BC-013, PG-BC-013

ID/CC	A 36-year-old female visits her family doctor because of anxiety, palpitations, **intolerance to heat**, nervousness with **trembling hands**, and **weight loss** despite a normal appetite.
HPI	She is also concerned about increasing **protrusion of her eyes** (EXOPHTHALMOS).
PE	VS: **tachycardia**; hypertension (BP 150/80). PE: wide pulse pressure; sweaty palms; warm skin; exophthalmos (due to enlargement of extraocular muscles); **generalized enlargement of thyroid gland** with bruit (DIFFUSE GOITER); nodular lesions over anterior aspect of lower legs (PRETIBIAL MYXEDEMA).
Labs	**Markedly decreased TSH** (due to negative feedback of autonomously secreted thyroid hormones); **increased T_3, T_4, and free T_4 index**; positive TSH receptor antibodies and antinuclear antibodies; hypercalcemia. CBC: anemia.
Imaging	Nuc: increased radioactive iodine uptake measurement; enlarged gland.
Gross Pathology	Increased vascularity of thyroid gland with symmetrical enlargement.
Micro Pathology	Thyroid gland hypertrophy and hyperplasia; reduced thyroid hormone storage and colloid; infiltrative ophthalmopathy.
Treatment	Antithyroid drugs; radioactive iodine.
Discussion	Also called diffuse toxic goiter, Graves' disease is the most common cause of hyperthyroidism. It is idiopathic in nature but has an **autoimmune basis** and is associated with HLA-B8 and HLA-DR3. LATS, an IgG, is responsible for some manifestations. Signs and symptoms are due to excess circulating thyroid hormone.
Atlas Links	UCV1 PM-BC-014 UCV2 MC-090A, MC-090B

HYPERTHYROIDISM (GRAVES' DISEASE)

ID/CC A 41-year-old obese female comes to the ER with **severe epigastric pain radiating to the back** accompanied by nausea and vomiting; she had been advised to undergo laparoscopic removal of symptomatic **small gallbladder stones**.

HPI She was admitted to the surgical floor and treated for pancreatitis. On the third day, she developed **numbness of the fingers and around the mouth and tongue as well as painful leg cramps** (HYPOCALCEMIC TETANY).

PE VS: hypotension; tachycardia; fever. PE: dehydrated and in acute distress; bilateral basal hypoventilation; abdomen tender in epimesogastrium; hypocalcemic signs present; abduction and flexion of foot when peroneal nerve is tapped (POSITIVE PERONEAL SIGN); hyperexcitability while using galvanic current (ERB'S SIGN); facial spasm on tapping over cheek (CHVOSTEK'S SIGN); carpal spasm seen with arterial occlusion by blood pressure cuff (TROUSSEAU'S SIGN).

Labs CBC: marked leukocytosis (17,000) with neutrophilia. **Amylase and lipase markedly elevated** (due to acute pancreatitis). ECG: **Q-T prolongation**. Markedly **reduced serum calcium**; normal serum albumin.

Imaging KUB: increase in gastrocolic space; **sentinel loop**. CXR: small left pleural effusion.

Gross Pathology Hemorrhagic pancreatitis with edema and areas of gray-white necrosis; intraperitoneal free hemorrhagic fluid; chalky-white fat necrosis (**saponification of calcium with lipids**).

Treatment Treat pancreatitis. IV calcium gluconate.

HYPOCALCEMIA FROM PANCREATITIS

ID/CC A 73-year-old female complains of **weakness, painful muscle cramps**, and **constipation**.

HPI She suffers from chronic congestive heart failure (CHF) that has been treated with **digoxin** and **furosemide**. She was also on oral potassium tablets but has discontinued them because of gastric upset.

PE VS: irregularly irregular pulse (atrial fibrillation); hypertension (BP 145/90); no fever. PE: well hydrated; conjunctiva normal; jugular venous pulse slightly increased, S3 heard; mild hepatomegaly and pitting edema of lower legs (all due to CHF); deep tendon **reflexes hypoactive**.

Labs CBC: normal. Lytes: **hypokalemia**. ECG: flattening of S-T segment and T waves; **prominent U waves**.

Treatment Potassium-rich foods (chick peas, bananas, papaya); oral potassium supplements; gastric mucosal protective agents; magnesium supplements (deficiency of magnesium frequently coexists). Potassium-sparing diuretics.

Discussion Potent diuretics such as furosemide frequently cause excessive renal loss of potassium with symptomatic hypokalemia which, if severe, may be life-threatening. **In patients on digoxin, hypokalemia greatly increases toxicity**.

ID/CC A 48-year-old female who has been **on total parenteral nutrition** for 2 weeks complains of **weakness**, cramps, **palpitations, tremors, and depression**.

HPI One week ago, she underwent her fifth major abdominal operation for intestinal fistula and sepsis.

PE VS: **tachycardia**; hypotension. PE: patient looks **confused** and "run down"; agitation with muscular **spasticity and hyperreflexia**; heart sounds disclose skipped beats; mild hypoaeration at lung bases; abdomen with three colostomy bags at site of fistula; no peritoneal irritation; no surgical wound infection.

Labs CBC: neutrophilic leukocytosis. Lytes: **hypomagnesemia** (< 0.8 mmol/L); borderline hypokalemia; **hypocalcemia** (severe hypomagnesemia reduces PTH secretion); low 24-hour urinary magnesium excretion. ECG: **prolonged P-R and Q-T intervals; wide QRS; tall T waves; premature ventricular ectopic contractions**.

Treatment Magnesium supplementation. Hypokalemia and hypocalcemia resolve with magnesium replacement.

Discussion Homeostasis of magnesium is achieved through a balance between intestinal (small bowel) absorption and urinary excretion. Deficiency is associated with the use of a large amount of IV fluids, alcoholism, intestinal malabsorption or diarrhea, inadequate replacement in parenteral nutrition, kwashiorkor or marasmus, prolonged GI suction, intestinal fistula, renal tubular acidosis, and use of drugs such as diuretics, cisplatin, methotrexate, amphotericin B, cyclosporine and aminoglycosides.

ID/CC   A 37-year-old white female complains of **nausea, vomiting, and headache** on her first postoperative day; the charge nurse found her having a grand mal seizure.

HPI   She had **elective surgery** for a benign left ovarian cyst. Her medical history discloses no previous illness.

PE   VS: no fever; normal heart rate. PE: well hydrated; slight **confusion and lethargy** as well as general **weakness**; slight increase in JVP; no bleeding or dehiscence (opening of surgical wound) or infection from surgical wound; no peritoneal signs; significant bilateral lower extremity edema.

Labs   CBC: normal. Lytes: hyponatremia (Na 115). Remainder of routine lab exams normal; normal cortisol (done to exclude possible adrenal insufficiency); serum osmolality < 280.

Treatment   For hyper- or isovolemic hyponatremia: water restriction (with caution to avoid osmotic central pontine myelinolysis syndrome, which can occur while restoring sodium levels too quickly). For hypovolemic hyponatremia: isotonic saline, slowly.

Discussion   Hyponatremia is the most common electrolyte disturbance seen in hospitalized patients and is often iatrogenic in nature. In a postoperative setting, the metabolic response to trauma is to increase secretion of ADH, among other hormones, which, coupled with overzealous IV administration of hypotonic fluids, may lead to symptomatic hyponatremia.

HYPONATREMIA

ID/CC A 48-year-old obese white female who works as a janitor is brought to the ER **in a coma** after being found on the floor of her room.

HPI Her husband reveals that she has been having **episodes of early-morning dizziness** and confusion associated with **hunger** and walking; he adds that these **symptoms disappear after eating**. He also states that the patient has frequently been **nervous and irritable**.

PE VS: tachycardia (HR 105); BP normal. PE: patient comatose; mild skin pallor; **cold, sweaty hands**; no focal neurologic deficits; heart sounds rhythmic; no murmurs; lungs clear; abdomen soft; no masses; peristalsis present.

Labs Normal hemoglobin (14.4 mg/dL); BUN and creatinine normal. Lytes: normal. **Hypoglycemia** (blood glucose 38 mg/dL); elevated insulin; **elevated plasma immunoreactive C-peptide** (vs. exogenous insulin administration, where C-peptide is low). Positive 72 hour fasting test (increased insulin levels despite hypoglycemia).

Imaging CT: 1.5-cm **mass in tail of pancreas**. Nuc: mass takes up octreotide.

Gross Pathology Single adenomatous mass.

Micro Pathology Findings according to type of islet cell involved.

Treatment Immediate IV glucose infusion; surgical resection.

Discussion The most common pancreatic islet cell tumor is β-cell insulinoma (usually benign). Other types include glucagonomas, somatostatinomas, gastrinomas (ZOLLINGER-ELLISON SYNDROME), and excessive VIP-secreting tumor (VERNER-MORRISON SYNDROME). Islet cell tumors may be seen in multiple endocrine neoplasia (MEN) syndromes.

ID/CC	An 18-year-old female is brought to a local clinic because she has never had a menstrual period (PRIMARY AMENORRHEA) and shows a **lack of breast development**.
HPI	She has a **cleft lip and palate**. On directed questioning, she reports a diminished sense of smell (HYPOSMIA).
PE	VS: normal. PE: left cleft lip and incomplete unilateral cleft palate; marked **hyposmia** on olfactory testing; heart and lung sounds within normal limits; no palpable mass in abdomen and pelvis; **no pubic or axillary hair; no breast tissue**.
Labs	CBC/Lytes: normal. LFTs normal; **decreased GnRH; low FSH and LH**.
Imaging	XR, skull: normal sella turcica. MR, brain: absent olfactory bulb(s).
Treatment	Gonadotropins.
Discussion	Kallmann's syndrome is an **X-linked** disorder characterized by deficiency of GnRH with a resulting decrease in FSH and LH levels, producing an isolated hypogonadotropic hypogonadism. It is typically associated with agenesis or hypoplasia of the olfactory bulbs, producing anosmia or hyposmia (lack of stimulus for GnRH production due to absent olfactory bulb catecholamine synthesis). More common in **men**.

ID/CC   A 2-year-old girl, the daughter of an African immigrant, is admitted to the pediatric ward owing to an **increase in abdominal girth** and **failure to thrive**.

HPI   She recently arrived in the United States from her home country. She was breast-fed until 1 year of age, at which time her mother ran out of milk. She is apathetic and irritable and has been having frequent episodes of diarrhea.

PE   **Height and weight in fifth percentile; skin and hair depigmentation**; thinning of hair; dry skin; hyperkeratosis on axillae and groin; hepatomegaly and **ascites**; generalized pitting **edema**; loss of muscle; lethargy.

Labs   CBC: anemia; lymphopenia. **Hypoalbuminemia** (normal in marasmus). Lytes: hypokalemia; hypomagnesemia.

Imaging   US/CT: fatty liver. KUB: pancreatic calcification (due to tropical pancreatitis). XR: delayed bone age.

Gross Pathology   **Fatty** infiltration of **liver**.

Micro Pathology   Intestinal mucosal atrophy with loss of brush border enzymes; atrophy of pancreatic islet cells; widespread fatty infiltration of liver.

Treatment   Restore acid-base and electrolyte balance; treat infections; gradually initiate high-protein diet with vitamins and minerals.

Discussion   Kwashiorkor is a form of malnutrition caused by **protein deprivation** with **normal total caloric intake**.

KWASHIORKOR

ID/CC	A 68-year-old obese male is rushed to the ER after he was found **unconscious** on the floor of his office.
HPI	He had been on medication for coronary artery disease.
PE	On admission, he is found to be in an acute state of tissue **hypoperfusion** (SHOCK) with a barely palpable pulse, hypothermia, and bradycardia. Immediate treatment for cardiac shock is begun.
Labs	ECG: acute anteroseptal myocardial infarction. **Increased serum lactate**; hyperphosphatemia. ABGs: **severe metabolic acidosis** (pH 7.27); bicarbonate 14 mEq/L (low). **Increased anion gap** (19) with no ketoacids; BUN and creatinine normal.
Treatment	Treat precipitating cause of acidosis; administer bicarbonate if pH is < 7.2. Treat shock.
Discussion	A state of increased levels of lactic acid in blood (LACTIC ACIDOSIS) may be due to a number of causes, including **shock** and **sepsis** (both of which increase lactic acid production due to hypoxia), methanol poisoning, metformin toxicity, and liver failure (due to failure of lactic acid to be removed from blood by its transformation to glucose). The anion gap is an estimation of the total unmeasured plasma anions, such as proteins, organic acids, phosphate, and sulfate. Increased anion-gap metabolic acidosis is due to salicylate poisoning, alcohol (e.g., methanol, ethanol, propylene glycol) intoxication, lactic acidosis, renal failure, and diabetic ketoacidosis.

LACTIC ACIDOSIS

ID/CC A 1-year-old female is taken to the emergency room because of **persistent vomiting** (20 times in 24 hours) that has been unresponsive to intramuscular antiemetics.

HPI While on a family vacation to Florida, she was given vanilla ice cream that was being sold on the street (dairy and meat products may harbor staphylococcal enterotoxins that produce food poisoning).

PE VS: tachycardia; mild fever; hypotension. PE. low urinary volume; **eyes sunken; poor skin turgor** with dryness of skin and mucous membranes; **lethargy and proximal muscle weakness** (due to hypokalemia).

Labs CBC: increased hematocrit (due to hemoconcentration); increased BUN. Lytes: **hypokalemia**; hypochloremia. UA: proteinuria; **high specific gravity**. ABGs: **metabolic alkalosis**. ECG: ST-segment and T-wave depression; U waves (hypokalemia).

Treatment Fluid and electrolyte replacement.

Discussion Dehydration may be isotonic, hypotonic, or hypertonic. When caused by protracted vomiting, it leads to metabolic alkalosis due to a decrease in hydrogen ion concentration with a compensatory rise in P_{CO_2} (due to diminished alveolar ventilation). The contraction of volume stimulates the proximal renal tubular cells to reabsorb bicarbonate in spite of alkalosis.

METABOLIC ALKALOSIS

ID/CC A 56-year-old man who is a known non-insulin-dependent **diabetic** (NIDDM **type II**, maturity onset) and who has been receiving an oral hypoglycemic agent is brought to the emergency room in a **stuporous state**.

HPI For approximately 2 weeks, he had been treated for a **URI** with oral antibiotics and bronchodilators.

PE VS: **tachycardia**; hypotension. PE: severe **dehydration** with dry oral mucosa and low urinary volume; patient **semiconscious** and **confused**; pupils react bilaterally and normally to light; evidence of proliferative diabetic retinopathy on funduscopic exam; no focal neurologic deficit found.

Labs CBC: mild leukocytosis (12,600). **Markedly increased blood glucose** (900 mg/dL); **increased serum and urinary osmolality** (> 350 mOsm/kg). Lytes: hypernatremia; mild hypokalemia. **Normal anion gap**. ABGs: normal serum bicarbonate (no acidosis). Elevated BUN and serum creatinine (suggestive of prerenal azotemia). UA: **glycosuria with no ketonuria**.

Treatment Hypotonic (one-half normal) saline. Insulin infusion (e.g., lower dose than in ketoacidosis). Potassium and phosphate supplement as needed.

Discussion Hyperosmolar, hyperglycemic nonketotic coma occurs mainly in older NIDDM patients and is usually associated with an episode of physical or mental stress (check for silent MI); it is not associated with ketosis or ketoacidosis. Volume depletion is severe (average fluid deficit 25% of total body water), and the mortality rate is high.

ID/CC A 44-year-old male is admitted to the orthopedic department because he sustained a **femoral neck fracture when he fell from a small stool**; the type and magnitude of the fracture are not compatible with the patient's age and impact.

HPI The patient recently emigrated from Somalia and states that he has been suffering from increasing **leg weakness** and persistent **lower back pain**.

PE VS: normal. PE: complete right femoral neck fracture; on palpation, **tenderness of lumbar vertebrae** and pelvic rim.

Labs Mild anemia (Hb 10 g/dL). Lytes: normal. **Increased alkaline phosphatase; decreased levels of 25-OH-D$_3$; hypocalcemia; hypophosphatemia**; increased PTH.

Imaging XR, hip: surgical neck femoral fracture. XR, lumbar spine: **collapse of lumbar vertebrae; generalized osteopenia; pseudofractures** (appearance of nondisplaced fractures representing local bone resorption).

Micro Pathology Excess osteoid but poor mineralization.

Treatment Vitamin D, calcium (and sometimes phosphate) supplements; surgical treatment of fracture, physiotherapy.

Discussion A poor diet in vitamin D and calcium, **lack of sunlight exposure, intestinal malabsorption, renal insufficiency**, or target organ resistance may lead to osteomalacia in the adult (or rickets in children), with defective calcification of osteoid.

OSTEOMALACIA

ID/CC A 40-year-old male visits his internist for an evaluation of sudden (PAROXYSMAL) **attacks of headache, perspiration, and anxiety**; attacks are precipitated by exercise, emotional stress, postural changes, and, at times, urination.

HPI **Very high blood pressure** has been recorded at the time of previous paroxysms. The patient has a good appetite but looks cachectic; blood pressure recorded between paroxysms is normal. The patient has no history suggestive of renal disease.

PE VS: **hypertension** (BP 180/120). PE: hypertensive retinopathy changes on funduscopic exam.

Labs **Elevated blood sugar** (due to increased catecholamines). Lytes: normal. **Increased** 24-hour urinary free **catecholamines** and **vanillylmandelic acid (VMA)** levels.

Imaging CT/MR: 5-cm left **adrenal mass**; very high signal on T2-weighted MR. Nuc: MIBG localizes to tumor and metastases.

Gross Pathology Encapsulated, **dusky-colored**, round tumor mass with compressed adrenal gland remnants at periphery and foci of necrosis and hemorrhage.

Micro Pathology Nests of pleomorphic large cells with basophilic cytoplasm and **chrome-staining granules** in vascular stroma; argentaffin stains positive; membrane-bound secretory granules on electron microscopy.

Treatment Treat hypertensive crises with pharmacologic alpha and beta blockade (pretreat with alpha blockers prior to beta blockers); resection of tumor.

Discussion Pheochromocytoma is the most common **tumor of the adrenal medulla** in adults; its symptoms are produced by **increased production of catecholamines**. Of these tumors, 10% are extra-adrenal, 10% bilateral, 10% malignant, and 10% familial; 10% occur in children, and 10% calcify. May be associated with multiple endocrine neoplasia (MEN) IIA or IIB syndromes.

Atlas Links ⬜ⓊⒸⓋⒾ PM-BC-026, PG-BC-026A, PG-BC-026B

ID/CC	A **9-year-old** female is brought to her pediatrician because of **breast enlargement**.
HPI	Her mother also reports **cyclical vaginal bleeding** and the appearance of **pubic and axillary hair** since the age of 4; an older cousin developed similar signs and symptoms.
PE	Fully developed breasts; axillary and pubic hair present; normal mental development; height and weight greater than average for her age; no focal neurologic signs.
Labs	**Increased plasma FSH, LH**, and estradiol; pubertal pattern of increased gonadotropins after infusion of GnRH.
Imaging	XR: **advanced bone age**. US: ovary enlarged to pubertal size with cyst formation. CT/MR: no pituitary lesion.
Gross Pathology	**Ovarian cyst** formation (luteal); in idiopathic variety, no structural abnormality found.
Treatment	GnRH agonists; psychiatric support; continuous search for possible cause.
Discussion	The most common cause of precocious puberty is idiopathic or constitutional; less common causes include hypothalamic-pituitary tumors (pinealomas, hamartomas, gliomas) or lesions causing gonadotropin-dependent precocious puberty.

PRECOCIOUS PUBERTY

ID/CC	A 5-year-old **boy** is brought to the pediatrician because of intermittent **numbness and leg cramps**.
HPI	His father is also concerned about the fact that his child is **shorter** than his classmates.
PE	**Full, round face**; short neck; flat nasal bridge; right convergence squint and left **cataract; delayed dentition**; positive Chvostek's and Trousseau's signs.
Labs	CBC: normal. Lytes: **hypocalcemia** ($<$ 8.8 mg/dL); **hyperphosphatemia** ($>$ 5 mg/dL). **Increased** plasma **PTH; no increase** in renal **cAMP** and phosphate clearance **with PTH infusion**.
Imaging	XR: **fourth and fifth metacarpals** are **short; premature physeal closure**; thickening of cortices with demineralization.
Treatment	Calcitriol and calcium supplementation.
Discussion	Also called Seabright-Bantam syndrome, pseudohypoparathyroidism is an **X-linked dominant** disorder in which there is **resistance to PTH action on the renal tubule** and bone with resulting hypocalcemia. Two types exist according to the response of cAMP to PTH. In type I (as in this case), patients fail to exhibit a phosphaturic response or increased cAMP after administration of PTH. Type II is associated with Albright's hereditary osteodystrophy.

PSEUDOHYPOPARATHYROIDISM

ID/CC A **15-month-old Eskimo** boy is brought to the pediatric clinic by his parents because of **delayed dentition, poor growth and development**, frequent crying, and weakness.

HPI The infant's **diet** is **deficient in** eggs and **dairy products**, and he spends most of his time indoors (i.e., he has **no exposure to sunlight**).

PE Irritability; poor muscular development and muscle tone; abdominal distention; hypotonia of all muscles; anterior fontanelle open; **softening of occipital and parietal bones with elastic recoil** (CRANIOTABES); frontal bossing; enlargement **of costochondral junctions** (RACHITIC ROSARY); **bowing of legs; lineal chest depression along diaphragm** (HARRISON'S GROOVE).

Labs **Serum calcium normal or slightly low; decreased serum phosphorus**; increased alkaline phosphatase; low $1,25(OH)_2$-vitamin D level.

Imaging XR: **widening of growth plates**; osteopenia of cranial and long bones; irregularity and cupping of distal ends of long bones; pseudofractures in metaphysis (LOOSER'S LINES).

Gross Pathology Excess amount of **uncalcified bone** at junction of cartilage; bone stretched and pulled out of shape by gravity; increased osteoid seams; osteopenia; frontal bossing of skull; **pigeon breast deformity**.

Micro Pathology Defective mineralization of osteoid in epiphysis and diaphysis.

Treatment Increase calcium and vitamin D in diet.

Discussion Rickets is a disease of infancy and childhood involving **defective mineralization of osteoid** in bone skeleton and the neuromuscular system because of **low vitamin D** or calcium in the diet; it can also be due to low sunlight exposure (vitamin D conversion in skin) and chronic renal failure (BUN and phosphorus levels are high).

Atlas Link 󰄯󰄰󰄱󰄲 MC-308

ID/CC   A 61-year-old male smoker presents with headache, weakness, **fatigue**, and **decreased urinary output** (OLIGURIA).

HPI   He was recently diagnosed with **oat cell carcinoma of the lung**.

PE   Cardiac sounds normal; no murmurs; no arrhythmias; no pitting edema; no hepatomegaly; no jugular plethora (no evidence of cardiac disease); no asterixis, jaundice, spider nevi, or parotid enlargement (no evidence of hepatic disease).

Labs   **Decreased serum sodium** (HYPONATREMIA); **decreased serum osmolality** (< 280 mOsm/kg); normal or low BUN and serum creatinine; no proteinuria (no renal disease); adrenal and thyroid function tests normal. UA: **urine osmolality markedly increased** (versus psychogenic polydipsia where osmolality is decreased); hypernatriuria (urinary Na > 20 mEq/L). **Diminished blood uric acid level** (HYPOURICEMIA).

Treatment   **Water restriction** plus a high-salt diet. **Demeclocycline**.

Discussion   **Syndrome of inappropriate** (increased) secretion of **antidiuretic hormone** (SIADH) occurs with either increased hypothalamic secretion (e.g., CNS disease, postoperative states) or ectopic secretion (e.g., tumors such as oat cell carcinoma of the lung). There may also be increased sensitivity to the effect of ADH (as occurs with chlorpropamide, fluoxetine, and carbamazepine).

ID/CC A 29-year-old female is brought by ambulance to the emergency room from her workplace owing to **confusion, agitation**, diarrhea, and **vomiting**.

HPI Her sister has myasthenia gravis. She gives a history of **recent weight loss** (7 kg) and a **recent severe URI**.

PE VS: **fever** (39.3°C); **tachycardia** with irregular pulse; hypotension (BP 100/50). PE: irritability; **delirium**; exophthalmos; diffuse increase in size of thyroid gland (GOITER); lungs clear; abdomen soft and nontender; no masses; no peritoneal irritation; deep tendon reflexes brisk; no neck stiffness or focal neurologic signs.

Labs CBC/Lytes: normal. LP: CSF values normal. ECG: atrial fibrillation. **Elevated T_4, free T_4, and T_3; low TSH**.

Treatment Treatment involves inhibition of thyroid hormone synthesis (with propylthiouracil or methimazole); inhibition of stored thyroid hormone (with iodide and corticosteroids); suppression of the peripheral effects of thyroid hormone (with propranolol); digitalization of patients with CHF and atrial fibrillation; acetaminophen for fever; and treatment of precipitating factors (e.g., antibiotics for infections).

Discussion Thyroid storm, a medical emergency, is usually precipitated by surgical or medical stress (e.g., infection) placed on untreated or undertreated hyperthyroid patients. Prevention of postoperative crises is effected through use of iodine and antithyroid drugs.

THYROID STORM

ID/CC A 27-year-old Cuban political dissident visits a medical clinic complaining of **diminished visual acuity, primarily at night**.

HPI He recently arrived in the United States by boat after spending several years in prison.

PE VS: normal. PE: conjunctiva shows diminished tear lubrication with dryness (XEROSIS; when localized, it forms patches known as Bitot's spots) as well as keratinization and small corneal ulcers (XEROPHTHALMIA).

Micro Pathology Keratinizing metaplasia in conjunctiva; follicular hyperkeratosis with glandular plugging.

Treatment Vitamin A supplementation.

Discussion Vitamin A (RETINOL) is a fat-soluble vitamin derived from β-carotenes that is used for the synthesis of rhodopsin in the retina, for wound healing, and for epithelial cell growth and differentiation. Night blindness (NYCTALOPIA) is an early symptom of vitamin A deficiency; conjunctival xerosis and Bitot's spots are early signs. Corneal ulcers may progress to erosions and eventual destruction of cornea (KERATOMALACIA).

VITAMIN A DEFICIENCY

ID/CC	A 36-year-old black male who is known to be an **alcoholic** comes to the emergency room with shortness of breath, confusion, **foot drop**, and swelling of his legs.
HPI	He admits to getting drunk at least three times a week. His **diet** consists mainly of canned soup and cheap "junk food" that he eats during the periods in which he is not drunk.
PE	VS: tachycardia. PE: dyspnea; jugular venous distention; **extremities warm** to touch; cardiomegaly; hepatomegaly; 2+ pitting edema of both lower extremities; confusion with **nystagmus**; decreased deep tendon reflexes.
Labs	**Increased** RBC **transketolase activity** coefficient; low serum and urine thiamine levels.
Imaging	CXR: cardiomegaly with basal lung congestion.
Gross Pathology	Wernicke's encephalopathy shows congestion, hemorrhages, and necrosis in thalamus, hypothalamus (**mammillary bodies**), and paraventricular regions.
Micro Pathology	Demyelinization of peripheral nerves with axonal degeneration and fragmentation.
Treatment	**Thiamine**. Before administering glucose to an alcoholic, thiamine should be given to prevent encephalopathy (due to depletion of thiamine in glycolysis pathways). Alcoholics should also receive IV or oral folate and multivitamins.
Discussion	Lack of thiamine produces Wernicke-Korsakoff syndrome as well as **high-output heart failure (wet beriberi) and polyneuropathy (dry beriberi)**. Thiamine pyrophosphate (TPP) is a cofactor for the Krebs cycle enzymes α-ketoglutarate dehydrogenase and pyruvate dehydrogenase as well as transketolase (pentose phosphate pathway).

33 **VITAMIN B$_1$ DEFICIENCY (BERIBERI)**

ID/CC A 45-year-old **alcoholic** Hispanic male who recently underwent a strangulated hernia repair becomes irritable and weak, suffers significant weight loss, and develops a **rash** on his face, his neck, and the dorsum of his hands; he also suffers from **diarrhea** and **altered mental status**.

HPI After his operation (which involved a 5-cm small bowel resection), the patient became torpid and anorexic with lack of proper return of bowel function for about 3 weeks. His **diet** had been based on **corn** products.

PE Erythematous, nonpruritic, hyperpigmented, scaling rash of face, neck (CASAL'S NECKLACE), and dorsum of hands; angular stomatitis (CHEILOSIS) and glossitis; diminished touch and pain sensation in all four extremities; apathy, confusion, and disorientation.

Labs UA: low levels of urinary N-methylnicotinamide.

Micro Pathology Atrophy and ulceration of gastric and intestinal mucosa; posterior columns show neuronal degeneration and demyelination.

Treatment Oral nicotinamide.

Discussion Vitamin B_3 (NIACIN) deficiency (PELLAGRA) is commonly seen in alcoholics and is less frequently seen in patients with GI disorders or in elderly patients. In patients with carcinoid syndrome, tryptophan, the precursor of niacin, is used up to form serotonin. It is usually accompanied by other B vitamin deficiencies. The typical observed triad consists of **dermatitis, dementia, and diarrhea**.

ID/CC   A 9-month-old white female is brought to the pediatric clinic because of **listlessness and anorexia**.

HPI   She is the daughter of an unemployed **poor** urban couple and has never before seen a pediatrician or taken any medication. Her parents report a diet of **unsupplemented cow's milk**.

PE   Weakness; pallor; **hyperkeratosis** and **hemorrhagic perifolliculitis** of skin of lower extremities, forearms, and abdomen; purpuric skin rashes; **splinter hemorrhages** in nail beds of hands; tenderness and swelling of distal femur and costochondral junctions; **bleeding gums; petechiae** seen over nasal and oral mucosa.

Labs   CBC: microcytic, hypochromic anemia; leukopenia. Plasma and platelet levels of ascorbic acid low; **prolonged bleeding time**.

Imaging   XR: subperiosteal hemorrhages; both legs and knees show "ground glass" appearance of bones and epiphyses.

Gross Pathology   Growing bone shows diminished osteoid formation, hemarthrosis, and subperiosteal and periarticular hemorrhage; **defective collagen** (vitamin C hydroxylates proline and lysine); endochondral bone formation ceases (osteoblasts fail to form osteoid); existing trabeculae brittle and susceptible to fracture.

Treatment   Oral ascorbic acid (high doses may produce oxalate and uric acid stones).

Discussion   Vitamin C (ASCORBIC ACID) deficiency, or scurvy, is observed in smokers, oncologic patients, alcoholics, infants, and the elderly.

VITAMIN C DEFICIENCY (SCURVY)

ID/CC	A 47-year-old homeless **alcoholic** man (with a diet deficient in leafy vegetables) comes into the emergency room with weakness, **bleeding** gums, swelling in his right knee due to blood collection (HEMARTHROSIS), and **bloody vomit** (HEMATEMESIS).
HPI	The patient's diet consists of one meal a day of leftovers from fast-food restaurants. He was **given ampicillin** for diarrhea **2 weeks ago** (leading to suppression of vitamin K synthesis by colonic bacteria).
PE	Thin and **malnourished** with poor hygiene; conjunctival and nail bed pallor; **subcutaneous ecchymosis** in arms and legs; right knee **hemarthrosis**.
Labs	Anemia (Hb 9.7); **prolonged PT and PTT**; normal platelet count, fibrinogen level, and thrombin time.
Treatment	Vitamin K supplementation.
Discussion	Coagulation factors II, VII, IX, and X are dependent on vitamin K for their activity (through γ-carboxylation). Broad-spectrum antibiotic use, malabsorption, and lack of dietary vitamin K result in deficiency, manifested as bleeding. Since these factors are made by the liver, severe liver disease can cause coagulopathy.

ID/CC A 47-year-old male high-school teacher visits his internist because of **chronic watery diarrhea** and **hot flashes while drinking alcohol**; a few months ago he also noticed a peculiar **redness of his face**.

HPI Every time he works or exercises in the sun, he develops a rash on exposed areas (PHOTODERMATITIS).

PE VS: normal. PE: in no acute distress; **redness of face**; no neck masses or increased JVP; **systolic ejection murmur grade I/IV at pulmonary area**, increasing with inspiration (pulmonary stenosis); wheezing heard; abdomen soft and nontender; mild **hepatomegaly**.

Labs CBC/Lytes: normal. Glucose, BUN, creatinine, and LFTs normal; no ova or parasites in stool. UA: **increased 5-hydroxyindoleacetic acid (5-HIAA)** in urine (a product of serotonin degradation).

Imaging KUB: ladder-step air-fluid levels. UGI: small bowel loops kinked, causing obstruction. CT: starlike thickening of mesentery due to desmoplastic retraction; vague liver metastatic lesions.

Gross Pathology Firm, yellow, submucosal nodule in a segment of ileum.

Micro Pathology Argentophilic cells (KULCHITSKY CELLS) in the intestinal crypts of Lieberkühn invading into mesentery; marked fibrotic reaction.

Treatment Octreotide, cyproheptadine (SOMATOSTATIN ANALOG).

Discussion Carcinoid tumors arise from the gastrointestinal tract or bronchi. These tumors secrete **serotonin** (5-HYDROXYTRYPTAMINE), producing the typical clinical syndrome. There may be stenosis of the pulmonic and tricuspid valve and **right-sided heart failure**.

CARCINOID SYNDROME

ID/CC A 3-week-old male is seen by a neonatologist because of **severe jaundice** that appeared at birth and has been worsening ever since.

HPI He is the first-born child of a healthy **Jewish** couple. His mother had an uneventful pregnancy and delivery.

PE Average weight and height for age; in no acute distress; **marked jaundice** (jaundice appears at levels of bilirubin around 2.5 to 3.0 mg/dL); slight hepatomegaly.

Labs **Markedly increased serum unconjugated bilirubin** (15 mg/dL); **very low fecal urobilinogen.**

Treatment Phenobarbital for type II. For type I, the prognosis is guarded, with the likelihood of death in the first year of life.

Discussion Crigler–Najjar syndrome is an inherited disorder of bilirubin metabolism that is characterized by a **deficiency** of the enzyme **glucuronyl transferase** and hence by an inability to conjugate bilirubin, with accumulation of indirect bilirubin and risk of kernicterus with brain damage (at bilirubin concentrations of > 20 mg/dL). There are two types: type I, which is more severe and is autosomal recessive, and type II, which is autosomal dominant.

ID/CC A 21-year-old female college student visits her gastroenterologist for an evaluation of fatigability and intermittent **right upper quadrant and epigastric pain**.

HPI She asked her family doctor to refer her to a gastroenterologist because she was concerned about her pain despite her doctor's reassurance that it was "nothing important."

PE VS: normal. PE: mild **jaundice** in conjunctiva and underneath tongue; well hydrated and in no acute distress; no hepatosplenomegaly on abdominal exam; no signs of hepatic failure.

Labs **Increased direct bilirubin** (vs. Gilbert's syndrome, in which hyperbilirubinemia is indirect) **and indirect bilirubin; liver enzymes mildly elevated**. UA: bilirubin and urobilinogen (vs. Gilbert's syndrome); **ratio of coproporphyrin I and coproporphyrin III in urine 5:1** (normal = 1).

Imaging US: no gallstones; liver normal. Nuc: no biliary excretion on HIDA.

Gross Pathology **Liver** normal size and **dark green** in color (versus Rotor syndrome, which has no gross liver abnormalities); absence of gallbladder inflammation or stones.

Micro Pathology Centrilobular, lysosomal granules with brownish pigment (catecholamines).

Treatment Supportive.

Discussion Dubin–Johnson syndrome is a benign **autosomal-recessive** disorder (vs. Gilbert's syndrome) of **defective canalicular bilirubin excretion** characterized by episodes of intermittent jaundice.

DUBIN–JOHNSON SYNDROME

ID/CC A 19-year-old male with a URI visits his family doctor because he is concerned about **yellowness in his eyes** (JAUNDICE), which he has noticed **whenever he is fatigued** or is suffering from a minor infection.

HPI He has no history of dark-colored urine, clay-colored stools, abdominal pain, blood transfusions, or drug use. He is immunized against hepatitis B and does not drink alcohol.

PE **Normal except for mild scleral icterus**; no hepatosplenomegaly; no signs of chronic liver failure.

Labs Moderately **increased serum bilirubin, predominantly unconjugated**; normal serum transaminases and alkaline phosphatase; normal serum albumin; **serum bilirubin rises after 24-hour fast**.

Treatment No metabolic treatment available or necessary.

Discussion The most common example of idiopathic hyperbilirubinemia is Gilbert's disease, which is autosomal dominant with variable penetrance. It is due to **defective bilirubin uptake by liver cells** and **low glucuronyl transferase activity**. Bilirubin levels seldom exceed 5 mg/dL, mainly unconjugated, and may vary inversely with caloric intake.

ID/CC   A 19-year-old female is brought to her family doctor by her parents, who have noticed that she has started **behaving oddly**; 2 days ago they noticed that her **eyes were yellow**.

HPI   She also complains of **tremor of her hands at rest** and some **rigidity** when trying to grasp objects (basal ganglia affectation). The **parents** of the patient are **first cousins**.

PE   Patient shows flapping tremor (ASTERIXIS) of hands; slit-lamp examination reveals **copper deposits in Descemet's membrane** of the cornea (KAYSER–FLEISCHER RINGS); abdominal palpation shows moderate **splenomegaly**.

Labs   CBC: hemolytic anemia (due to oxidative RBC damage by copper). **AST and ALT elevated** as well as alkaline phosphatase and bilirubin, both direct and indirect; **decrease in serum ceruloplasmin** (copper-transporting protein); **increased urinary copper** (HYPERCUPRIURIA); increased urinary uric acid (HYPERURICOSURIA).

Imaging   US: enlargement of liver and spleen.

Gross Pathology   Copper accumulation in liver, brain, and cornea.

Micro Pathology   Liver biopsy shows acute inflammation, increased copper levels, and periportal fibrosis (macronodular cirrhosis); **intracytoplasmic hyaline bodies** (MALLORY BODIES) seen; **degeneration of basal ganglia** with cavitation, especially of putamen; hyperplasia with glial proliferation of the lenticular nuclei.

Treatment   **Penicillamine** (copper chelating drug), pyridoxine. Consider liver transplantation.

Discussion   Wilson's disease is an **autosomal-recessive** inherited disorder of copper metabolism mapped to chromosome 13. It is characterized by **increased absorption of copper from the intestine and diminished excretion in the bile** with resultant copper deposition, primarily in the brain and liver.

WILSON'S DISEASE

ID/CC A 23-year-old white **female** is brought to the ER because of strange, dreamlike **hallucinations and blurred vision** that she experienced 1 day **after spending all morning in the sun** painting her house (exposure to sun may precipitate attacks).

HPI The patient had undergone two **previous laparotomies** for apparent acute abdomen, but **no pathology was found**. She has had several episodes of **recurrent abdominal pain**.

PE VS: no fever or tachycardia. PE: pupils are of unequal size (ANISOCORIA); generalized weakness and hypoactive deep-tendon reflexes; disorientation; **foot drop; urine very dark** and foul-smelling. No photosensitive skin lesions.

Labs UA: **increased urine porphobilinogen** and **γ-aminolevulinic acid**. Lytes: hyponatremia.

Gross Pathology Liver infiltrated with porphobilinogen; central and peripheral nervous system myelin sheath degeneration.

Micro Pathology Degeneration of myelin sheath.

Treatment High-carbohydrate diet; glucose; hematin.

Discussion Acute intermittent porphyria is an **autosomal-dominant deficiency in** an enzyme of porphyrin metabolism **(porphobilinogen deaminase)** that leads to systemic symptoms, acute abdominal pain, neuropsychiatric signs and symptoms, and CNS and peripheral neuropathy. Acute intermittent porphyria is differentiated from other porphyrias by its **lack of photosensitive skin lesions**. Sun exposure and drugs (e.g., sulfa, barbiturates) may precipitate attacks.

ACUTE INTERMITTENT PORPHYRIA

ID/CC	A 27-year-old **farmer from Florida** (with abundant sun exposure) comes to see his dermatologist for an evaluation of a recent **increase in size and change in color of a skin lesion** that has been present on the dorsum of his hand (a sun-exposed area) for 6 years.
HPI	The patient is an **albino**, but he has not been able to comply with his dermatologist's orders to wear long sleeves while working in the field.
PE	**White hair, including eyelashes and eyebrows**; eye exam shows **nystagmus** and poor development of macula with blue iris; poor visual acuity (20/350); skin is pink-white with lack of pigmentation throughout body; numerous actinic (SOLAR) keratoses on face and scalp as well as on dorsum of hands; **ulcerated lesion with indurated edges** on dorsum of hand with hyperpigmentation.
Labs	**Tyrosine** assay shows **absence** of the enzyme.
Gross Pathology	Patches of scaly, irregular, hypertrophied skin in sun-exposed areas **(actinic keratosis)**.
Micro Pathology	Biopsy of lesion on dorsum of hand shows epidermoid **(squamous cell)** cancer with epithelial pearls.
Treatment	Surgery and/or chemotherapy for skin cancers, avoidance of sun exposure, management of visual impairment.
Discussion	Albinism is a hereditary disorder that may be generalized or localized and is transmitted as an autosomal-dominant or autosomal-recessive trait. It is always distinguished by various degrees of **hypopigmentation** of the skin, hair, iris, and retina. The defect lies in the pigmentation, not in the number of melanocytes present in the body. The cause is an **absence of tyrosinase**, the enzyme that catalyzes the conversion of tyrosine to dihydroxyphenylalanine and melanin. There is a marked increase in the risk of skin cancer.

ID/CC A 37-year-old man presents with **dark, blackened spots in his sclera and ear cartilage** as well as **back pain** and restriction of motion with **pain and swelling of both knee joints**.

HPI Directed questioning reveals that his **urine turns black** if left standing.

PE **Increased pigmentation** in ears, conjunctiva, nasal bridge, neck, and anterior thorax (OCHRONOSIS); arthritis of spine, both knee joints, and fingers.

Labs UA: **elevated urine homogentisic acid** (causes urine to darken upon standing or with addition of alkaline substances).

Imaging XR: calcification in cartilage of knee menisci and wrist; premature arthritic changes.

Treatment Symptomatic treatment of arthritis.

Discussion Also called **ochronosis**, alkaptonuria is an autosomal-recessive disorder of tyrosine metabolism characterized by the **absence of homogentisate oxidase** due to a defective gene on chromosome 3 with accumulation of homogentisic acid in cartilage, giving a dark blue discoloration to the tissues and leading to degenerative joint disease.

ID/CC	An **11-year-old white** female is brought to the ER by her parents because of fever, **difficulty breathing, and a productive cough with greenish sputum.**
HPI	Her parents are of northern European descent. She has a history of **recurrent UTIs and foul-smelling diarrhea** since infancy.
PE	VS: tachycardia; tachypnea (RR 45). PE: mild cyanosis; malnourishment; **nasal polyps**; hyperresonance to lung percussion with **barrel-shaped chest**; scattered rales; hepatomegaly.
Labs	**High sodium and chloride** concentrations in **sweat test**; *Pseudomonas aeruginosa, Haemophilus influenzae* and *Staphylococcus aureus* in sputum culture. PFTs: increased RV/TLC ratio. Increased **fecal fat**. ABGs: hypoxemia; hypercapnia.
Imaging	CXR: few dilated bronchi (BRONCHIECTASIS) filled with mucus; emphysema; XR, paranasal sinuses: opacification of sinuses.
Gross Pathology	Atrophic pancreas with almost complete disruption of acini and replacement of exocrine pancreas with fibrous tissue and fat; mucous plugging of canaliculi.
Micro Pathology	Inflammatory change.
Treatment	Antibiotics, low-fat diet, and supportive measures; recombinant human DNase (cleaves extracellular DNA from neutrophils in sputum); inhaled amiloride, nasal corticosteroids, systemic corticosteroids, decongestants; consider lung transplant.
Discussion	Cystic fibrosis is an **autosomal-recessive** disease that is due to a mutation in the long arm of **chromosome 7** (band q31) in the cystic fibrosis transmembrane conductance regulator (CFTR) gene. If CFTR function is deficient, **chloride** and water transport is slowed and secretions are inspissated.

CYSTIC FIBROSIS

ID/CC A 15-year-old female is brought to the emergency room from school following the sudden development of **severe, intermittent right-flank pain** together with nausea, vomiting, and **blood in her urine** (a picture typical of renoureteral stone).

HPI Her medical and family history is unremarkable.

PE VS: tachycardia; normal BP; slight fever. PE: short stature (due to lysine deficiency); in acute distress; **constantly switches positions in bed** (due to renal colic); abdominal tenderness; no peritoneal irritation; costovertebral angle tenderness.

Labs **Increased urinary excretion of cysteine, ornithine, arginine, and lysine** on urine amino acid chromatography (due to intestinal and renal **defect in reabsorption**). UA: hematuria; hexagonal crystals (CYSTEINE) upon cooling of acidified urine sediment.

Imaging KUB/IVP/CT urography: radiopaque stone in area of right kidney.

Treatment **Low-methionine diet**; increase fluid intake; alkalinize urine; penicillamine.

Discussion Cystinuria is an **autosomal-recessive** disorder of dibasic amino acid metabolism (due to impaired renal tubular reabsorption); it leads to increased cysteine urinary excretion and **kidney stone formation**.

CYSTINURIA

ID/CC A 9-year-old boy is brought to the emergency room with pain, inability to move his left shoulder, and flattening of the normal rounded shoulder contour (SHOULDER DISLOCATION) that occurred when he tried to hit a ball with his bat at a local baseball field.

HPI He has **dislocated his left shoulder nine times before and his right shoulder three times before**. He also has a history of **easy bruising**.

PE Hyperelastic skin; **"cigarette paper"** scars in areas of trauma; **hyperextensibility of joints**; left shoulder dislocated; multiple bruises over skin.

Labs Clotting profile normal.

Imaging XR: left shoulder dislocated.

Micro Pathology Collagen fibrils of dermis of skin larger than normal and irregular in outline on electron microscopy.

Treatment Supportive.

Discussion Ehlers–Danlos syndrome is also known as cutis hyperelastica. **Faulty collagen synthesis** produces 10 types of Ehlers-Danlos syndrome, some of which are autosomal recessive (type VI), others autosomal dominant (type IV), and others associated with X-linked recessive transmission (type IX). Prone to aneurysm and dissection in the great vessels.

EHLERS–DANLOS SYNDROME

ID/CC   A 17-year-old **male** presents with episodes of **painful, burning paresthesias along his palms and soles** along with markedly **diminished vision** in his right eye.

HPI   His maternal **uncle died of chronic renal failure** at the age of 40.

PE   Clusters of **purplish-red, hyperkeratotic lesions** on skin around umbilicus, buttocks, and scrotum (ANGIOKERATOMAS); **right corneal leukomatous opacity**; neurologic exam normal except for painful paresthesias along arms and soles; pitting edema in lower extremities.

Labs   **Elevated serum creatinine and BUN** (patients usually die of renal failure). UA: proteinuria; broad casts. PBS: leukocytes reveal deficiency of α-galactosidase.

Micro Pathology   Lipid deposition in epithelial and endothelial cells of glomeruli and tubules (FOAM CELLS) on renal biopsy; **lysosomal accumulation of glycosphingolipid (ceramide trihexoside)** in the form of "myelin bodies" on electron microscopy of skin, heart, kidneys, and CNS.

Treatment   Treat pain crises symptomatically; renal failure may require renal transplantation.

Discussion   Fabry's disease, a sphingolipidosis, is a **rare X-linked recessive disorder** of glycosphingolipid metabolism caused by a **deficiency of α-galactosidase A** and by the consequent accumulation of ceramide trihexoside.

ID/CC	A **28-year-old** white male complains of severe **retrosternal pain** radiating to his left arm and jaw.
HPI	He has not had a physical exam in a long time. He adds that his **father died at a young age of a myocardial infarction**.
PE	Anguished, dyspneic, diaphoretic male with hand clutched to chest (indirect sign of myocardial pain); soft, **elevated plaques on eyelids** (XANTHELASMAS); arcus senilis; painful **xanthomas of Achilles tendons** and patellae.
Labs	Elevated CK-MB; elevated troponin T and I. ECG: MI. Extremely **high levels of LDL**.
Imaging	Angio: coronary artery disease.
Gross Pathology	**Premature atherosclerosis** in large arteries.
Micro Pathology	Foam cells with lipid characteristic of atherosclerotic plaques.
Treatment	Diet, exercise and cholesterol-lowering drugs (although HMG-CoA reductase inhibitors are ineffective in homozygous FH patients due to complete lack of LDL receptors). Consider portocaval anastamosis or liver transplantation.
Discussion	Familial hypercholesterolemia is also called type II hyper-lipoproteinemia; it is an **autosomal-dominant defect in LDL receptor** with a gene frequency of 1:500. Homozygotes may have an LDL count eight times that of normal.

GENETICS

FAMILIAL HYPERCHOLESTEROLEMIA

ID/CC	A 16-year-old white female complains of sudden **midepigastric pain and nausea** after eating french fries.
HPI	Her history reveals that **she and a sibling** have had similar episodes of abdominal pain in the past. Careful questioning discloses that she experiences **flushing** every time she **drinks alcohol**.
PE	**Nonpainful, yellowish papules on face, scalp, elbows, knees, and buttocks** (ERUPTIVE XANTHOMATOSIS); lipemia retinalis on funduscopic exam; hepatosplenomegaly; abdominal muscle guarding and palpable tenderness.
Labs	**Elevated serum amylase** and lipase; **very high triglycerides**; moderate elevation of serum cholesterol and phospholipids.
Micro Pathology	Lipid-laden foam cells.
Treatment	**Low-fat diet**; avoidance of alcohol; exercise; **fibric acid** and **niacin** in selected cases.
Discussion	Familial hypertriglyceridemia is an autosomal-dominant disorder. Abdominal pain stems from **recurrent acute pancreatitis**.

ID/CC An 8-year-old boy is referred to the pediatric clinic for evaluation of **anemia and multiple developmental anomalies**.

HPI His parents report that he **bleeds easily**.

PE Pale and **mentally retarded;** small head (MICROCEPHALIA); low height and weight for age; hyperpigmentation of torso and thighs with café-au-lait spots; **decrease in size of penis**; decrease in size of eyes (MICROPHTHALMIA); **absence of both thumbs**.

Labs CBC: decreased WBCs (LEUKOPENIA), platelets (THROMBOCYTOPENIA), and RBCs (ANEMIA) (PANCYTOPENIA). Increased levels of HbF. Bone marrow chromosomes show diverse alterations (breaks, constrictions, and translocations).

Imaging XR: **bilateral absence of radii**. IVP/CT: **hypoplastic kidneys**.

Treatment **Marrow transplantation**, androgens, corticosteroids.

Discussion Fanconi's anemia is a congenital, autosomal-recessive disorder characterized by constitutional **aplastic anemia due to defective DNA repair**, presumably as a result of viral infection causing hypersensitivity to DNA cross linking agents. It is associated with multiple musculoskeletal and visceral anomalies, proximal renal tubular acidosis, and a higher incidence of leukemia.

FANCONI'S ANEMIA

ID/CC	A 10-year-old **male** is referred to a genetic evaluation clinic by his pediatrician because of **mental retardation**.
HPI	His mother did not take any drugs during her pregnancy, did not suffer from any major illnesses, was seen by an obstetrician periodically, and was monitored intrapartum.
PE	Patient well developed physically with grade I mental retardation; no evidence of cardiovascular, genitourinary, or hepatic disease.
Labs	Patient has been subjected to basic and endocrinologic lab profiles, all of which have yielded normal results. Karyotype: **"fragile gap" at end of the long arm on X chromosome**.
Treatment	Supportive.
Discussion	The **second most common cause of mental retardation** after Down's syndrome **in males** (women are carriers), fragile X syndrome should be suspected in any male patient whose mental retardation cannot be explained by other disease processes. It is often associated with macro-orchidism, large ears and jaw, a high-pitched voice, and connective tissue abnormalities and demonstrates genetic **anticipation** (worsening of the disorder in successive generations) owing to the expansion of **trinucleotide repeats**.

ID/CC	A 2-month-old white male is taken to his family doctor because of lethargy, **feeding difficulties**, and yellowish skin (JAUNDICE).
HPI	The child has been **vomiting** on and off since birth.
PE	Irritability; **jaundice; cataracts; hepatomegaly**; growth and development in fifth percentile; edema.
Labs	UA: **galactosuria**; aminoaciduria; albuminuria. **Hypoglycemia**; increased ALT and AST; elevated direct and indirect bilirubin; prolonged PT; erythrocytes have **markedly reduced galactose-1-phosphate uridyl transferase activity** and elevated galactose-1-phosphate.
Imaging	US/CT: enlarged fatty liver.
Gross Pathology	Early hepatomegaly and fatty change with giant cells leading to cirrhosis; gliosis of cerebral cortex, basal ganglia, and dentate nucleus of cerebellum; cataracts.
Micro Pathology	Liver, eyes, and brain most severely affected by **deposits of galactose-1-phosphate and galactitol**; kidney, heart, and spleen also involved.
Treatment	Limit intake of milk and other galactose- and lactose-containing foods.
Discussion	Galactosemia is an **autosomal-recessive** lack of enzyme galactose-1-phosphate uridyl transferase; the presence of **cataracts** differentiates it from other causes of jaundice in the newborn.

GALACTOSEMIA

ID/CC   An 11-year-old **Jewish** male presents with weakness, **epistaxis**, and a left-sided abdominal mass.

HPI   He has a history of **bruising easily** and sustaining fractures following minimal trauma.

PE   Mental retardation; multiple **purpuric patches**; skin pigmentation; mild hepatomegaly; **massive splenomegaly**; marked pallor; no lymphadenopathy or icterus.

Labs   CBC: normocytic, normochromic anemia; thrombocytopenia; low normal WBC count. LFTs normal; bone marrow biopsy characteristic; isolated WBCs demonstrate reduced β-glucosidase activity; elevated serum acid phosphatase.

Imaging   XR, spine: biconcave (H-shaped) vertebral bodies. XR, knee: Erlenmeyer flask deformity of distal femur; osteopenia. CT/US: enlarged spleen with multiple nodules.

Micro Pathology   Bone marrow biopsy shows myelophthisis; replaced by Gaucher's cells 20 to 100 μm in size; characteristic **"wrinkled paper" cytoplasm** due to intracytoplasmic glucocerebroside deposition; PAS stain positive.

Treatment   Symptomatic, enzyme replacement with purified placental or recombinant acid β-glucosidase, splenectomy.

Discussion   Gaucher's disease is an **autosomal-recessive deficiency of glucocerebrosidase** with accumulation of glucosyl-acylsphingosine in bone marrow, spleen, and liver.

ID/CC	A 5-month-old male is brought to the doctor because of frequent nausea, **vomiting**, night sweats, tremors, and **lethargy**.
HPI	When the patient was exclusively breast fed (i.e., during the initial four months after birth), he was well; the **onset of symptoms coincided with** the occasional **addition of fruit juices** to the baby's diet.
PE	Lazy-looking, slightly **jaundiced** baby; mild growth retardation; **hepatomegaly**.
Labs	**Marked hypoglycemia; fructosemia**. UA: fructosuria; urine test for reducing sugar positive; dipstick for glucose negative; fructose tolerance test not advisable (may cause severe hypoglycemia).
Micro Pathology	Liver biopsy reveals **low aldolase B activity** (confirmatory test).
Treatment	Return to breast feeding as sole food; avoid fruit juices, fruits, and sweets.
Discussion	Any food containing fructose or sucrose (fructose + glucose) may cause symptoms in patients with fructose intolerance, an **autosomal-recessive deficiency of aldolase B** (enzyme used to split fructose-1-phosphate into glyceraldehyde and dihydroxy-acetone phosphate), resulting in accumulation of fructose-1-phosphate within liver cells. This inhibits glycolysis, gluconeogenesis and glycogenolysis. If long-standing, it may lead to cirrhosis and kidney failure. Differential diagnosis is galactosemia.

GENETICS

HEREDITARY FRUCTOSE INTOLERANCE

ID/CC	A 9-year-old male is referred to the pediatric clinic because of progressive **mental retardation, diminished visual acuity**, and **bone deformity** in the thorax.
HPI	The boy was born in Malaysia and never had any prenatal screening.
PE	Tall and thin with elongated limbs (Marfanoid appearance); fine hair; **abnormally long fingers** (ARACHNODACTYLY); **pectus excavatum; lenticular dislocation** (ECTOPIA LENTIS); malar flush; high-arched palate; genu valgum; cardiovascular exam normal.
Labs	**Increased serum methionine; increased urinary homocystine**.
Imaging	XR: generalized **osteoporosis**.
Micro Pathology	Brain gliosis; fatty liver; arterial intimal thickening without lipid deposition; degeneration of zonular ligaments of lens.
Treatment	High-dose pyridoxine (cofactor for cystathionine synthetase; effective only in some forms of disease); methionine-restricted diet; cysteine and folate supplements.
Discussion	Homocystinuria is an autosomal-recessive **disturbance of methionine metabolism** caused by a **deficiency of cystathionine synthetase** in liver cells with accumulation of homocystine. Major arterial and venous **thromboses** are a constant threat because of vessel wall changes and increased platelet adhesiveness due to the toxicity of homocystine to the vascular endothelium.

HOMOCYSTINURIA

ID/CC	An 11-year-old **male** is sent to the audiometry clinic by his pediatrician for an evaluation of **deafness**.
HPI	His teachers note that he has not been paying attention at school and add that his academic performance has suffered as a result.
PE	**Coarse facies and large tongue**; short stature; **corneas clear** (vs. Hurler's disease); dimpled skin in back of arms and thighs; no gibbus (acute-angle kyphosis) present (vs. Hurler's disease); nonpainful nodular lesions on left scapular area; stiffening of joints; **deafness**.
Labs	UA: **increased urinary heparan sulfate and dermatan sulfate**.
Imaging	Metacarpal thickening with tapering at ends.
Micro Pathology	Metachromatic granules (REILLY BODIES) in bone marrow leukocytes; amniotic fluid culture during pregnancy may detect abnormality.
Treatment	Supportive.
Discussion	Hunter's disease, or type II mucopolysaccharidosis, is an **X-linked recessive** disease and is less severe than Hurler's syndrome (type I). Hunter's disease can be differentiated from Hurler's syndrome in that it features no corneal opacities and either no mental retardation or less severe retardation than that found in Hurler's; however, deafness is present. Caused by a **deficiency of iduronosulfate sulfatase**.

HUNTER'S DISEASE

ID/CC	A **2-year-old** white male is brought to the ophthalmologist for an evaluation of **eye clouding**.
HPI	The child has a physical and **mental disability** very similar to that of his older **brother**.
PE	**Short stature**; very **coarse, elongated facial features** (GAR-GOYLISM); **bilateral corneal opacities**; retinal degeneration and papilledema; saddle nose deformity; systolic murmur in second right intercostal space; **enlarged heart, liver, and spleen; kyphoscoliosis** with lumbar gibbus (acute angle kyphosis); stiff, immobile, and contracted large joints.
Labs	Dermatan sulfate and heparan sulfate in urine; **α-L-iduronidase deficiency in WBCs**.
Imaging	XR: dolichocephaly; increased diameter of sella turcica; deformation of vertebral bodies with scoliosis and kyphosis.
Gross Pathology	Increased mucopolysaccharide (MPS) deposition in heart, eye, connective tissue, CNS, cartilage, heart, and bone.
Micro Pathology	Enlarged heart; thickened endocardium; MPS infiltration in intima of coronary arteries; meningeal and neuronal deposits producing hydrocephalus; metachromatic granules in lymphocytes and histiocytes.
Treatment	Supportive ophthalmologic, skeletal, and cardiovascular treatment. Consider bone marrow transplant.
Discussion	Also known as gargoylism, Hurler's syndrome is the most common mucopolysaccharidosis (TYPE I). It is **autosomal recessive** and is caused by a **deficiency of α-iduronidase**. Death usually occurs by 6 to 10 years of age, usually secondary to cardiovascular complications.
Atlas Link	UCV2 Z-BC-058

ID/CC	A 25-year-old male visits a fertility clinic as part of an **evaluation of infertility** that he is undergoing with his wife.
HPI	His medical history discloses frequent **sinus infections** (SINUSITIS) and chronic cough with sputum formation (BRONCHIECTASIS).
PE	VS: normal. PE: apical impulse felt on fifth **right intercostal space**; all auscultatory foci reversed (DEXTROCARDIA); liver on left side and spleen on right (SITUS INVERSUS).
Labs	CBC/Lytes: normal. Semen analysis shows **immotile spermatozoa**.
Imaging	CXR: dextrocardia. KUB: situs inversus.
Discussion	Kartagener's syndrome (also called immotile cilia syndrome) is an **autosomal-recessive** disorder characterized by **lack of dynein** (ATPASE) arms from the microtubules of axonemes in the cilia of the sinuses and bronchi, rendering them immotile. Sperm are also immotile (due to flagellar lack of dynein). The lack of mucus-clearing action causes frequent infections.

GENETICS

KARTAGENER'S SYNDROME

ID/CC	A 19-year-old male visits his family physician because he is **embarrassed at having large breasts**.
HPI	He also complains of frequent headaches and **impotence**.
PE	**Tall, eunuchoid** body habitus; mild mental retardation; testes small and firm; breast enlargement (GYNECOMASTIA); female distribution of hair.
Labs	UA: increased urinary FSH; decreased 17-ketosteroid.
Imaging	XR: delayed physeal closure; short fourth metacarpal.
Gross Pathology	Testicular atrophy.
Micro Pathology	**Testicular fibrosis and hyalinization; lack of spermatogenesis**; Leydig's interstitial cells scarce and have foamy cytoplasmic change; **female sex chromatin bodies** (BARR BODIES) in cells.
Treatment	Testosterone.
Discussion	Also known as **testicular dysgenesis**, Klinefelter's syndrome is the most common cause of male hypogonadism. Alteration is due to the presence of three sex chromosomes (karyotype 47,XXY).

ID/CC A **5-month-old** child is brought to the pediatrician because of **growth retardation** and **difficulty feeding**.

HPI His parents note that the child has been **irritable** and "stiff" (SPASTICITY).

PE VS: normal. PE: patient **underdeveloped** for age; **reflexes hyperactive**; paravertebral muscles and hamstrings tense (RIGIDITY); maternal milk **sucking reflex weak** and punctuated by periods of regurgitation.

Labs Basic lab work within normal limits. LP: increased protein in CSF.

Gross Pathology Axonal and white-matter cerebral, cerebellar, and basal ganglia **demyelination**.

Micro Pathology Basophilic perivascular multinucleated globoid cells (MACROPHAGES) with cytoplasmic inclusion bodies consisting of cerebroside.

Treatment Poor prognosis, with death usually occurring rapidly.

Discussion Also called **globoid leukodystrophy**, Krabbe's disease is an **autosomal-recessive**, familial genetic disorder characterized by a **deficiency of** galactosylceramide **β-galactosidase**.

KRABBE'S DISEASE

ID/CC	A 2-year-old **male** is brought to the pediatrician by his mother because of repeated, **self-mutilating biting of his fingers and lips**; the patient's mother has also noticed abundant, **orange-colored "sand"** (uric acid crystals) **in the child's diapers**.
HPI	The mother reports that some months ago the child's urine was red, but she took no action at the time.
PE	Poor head control, difficulty walking, and difficulty maintaining an erect, seated position; **choreoathetoid movements**, spasticity, and **hyperreflexia** on neurologic exam.
Labs	**Hyperuricemia** (>10 mg/dL). UA: crystalluria; microscopic hematuria (due to renal calculi).
Imaging	XR: irregular amputation of several fingers.
Treatment	Allopurinol. Removal of primary teeth.
Discussion	Lesch–Nyhan syndrome is an **X-linked recessive** metabolic disease resulting from a deficiency of an enzyme of purine metabolism, **HGPRT**. If left untreated, patients develop full-blown **gouty arthritis** and urate nephropathy as well as subcutaneous tophaceous deposits. Compulsive, uncontrollable destructive behavior is typical of the disorder. Prenatal diagnosis is possible.

ID/CC	A **5-day-old** male presents with **seizures**, difficulty feeding, and vomiting; his mother reports a **peculiar, maple-sugar-like odor on his diapers**.
HPI	His mother had an unremarkable full-term vaginal delivery.
PE	VS: no fever. PE: full-term neonate with irregular respirations, **muscular rigidity** (SPASTICITY), and obtunded sensorium; fundus normal; peculiar odor in urine and sweat; when child's head support (hand) is suddenly withdrawn in supine position, patient fails to react with normal extension-abduction followed by flexion and adduction of arms (ABSENCE OF MORO REFLEX).
Labs	Hypoglycemia. ABGs: metabolic acidosis. **Marked elevation in blood and urine levels of the branched chain amino acids** leucine, isoleucine, alloisoleucine, and valine as well as decreased levels of alanine, threonine, and serine.
Gross Pathology	Edema of brain with gliosis and white matter **demyelination**.
Treatment	**Restricting intake of branched-chain amino acids from diet**; dialysis; thiamine supplementation.
Discussion	Maple syrup urine disease is an **autosomal-recessive** branched-chain α-ketoaciduria that results from defective oxidative decarboxylation of the branched-chain α-ketoacids. This decarboxylation is usually accomplished by a complex enzyme system (**α-ketodehydrogenase**) using thiamine as a coenzyme. A **deficiency** of this enzyme system causes urine to have the characteristic maple syrup odor and causes CNS symptoms in the first few weeks of life.

MAPLE SYRUP URINE DISEASE

ID/CC	A 3-year-old white male is brought to the pediatrician because of **increasing difficulty walking** due to **spasticity**.
HPI	The child had been developing normally up to now, and his medical history is unremarkable.
PE	**Difficulty climbing stairs; ataxia; wide-based gait**; extensor plantar response and hyperreflexia.
Labs	LP: increased protein in CSF (vs. cerebral palsy). Decreased peripheral nerve conduction velocity.
Imaging	MR, brain: demyelination.
Gross Pathology	Generalized **demyelination** (due to deficiency of arylsulfatase A interfering with normal metabolism of myelin lipids) with gliosis.
Micro Pathology	Toluidine blue staining shows brownish (METACHROMATIC) granules in oligodendrocytes and neurons of globus pallidus, thalamus, and dentate nucleus.
Treatment	Poor prognosis; patients become invalids within a few years and die before puberty.
Discussion	Metachromatic leukodystrophy is an **autosomal-recessive** disorder of sphingolipid metabolism that is due to a **deficiency in the enzyme arylsulfatase A** with accumulation of sulfatides in the central and peripheral nervous system as well as in the kidneys. Intrauterine diagnosis is possible.

METACHROMATIC LEUKODYSTROPHY

ID/CC	An 11-month-old **Jewish** male of **Ashkenazi** descent presents with globally delayed development and **diminished visual acuity**.
HPI	His parents feel that the baby is not acquiring new skills and that existing ones are regressing. They also feel that their child cannot see or hear properly.
PE	**Lymphadenopathy; hepatosplenomegaly; cherry-red spot on macula** on funduscopy; malnourished infant with protuberant abdomen; global developmental delay; hypoacusis.
Labs	CBC: mild normochromic, normocytic anemia.
Micro Pathology	Bone marrow biopsy reveals sphingomyelinase deficiency in cultured skin fibroblasts; characteristic "foam cells" containing sphingomyelin and cholesterol.
Treatment	No treatment available. Carries a poor prognosis, with death occurring within a few years of birth.
Discussion	Niemann–Pick disease is an **autosomal-recessive deficiency of sphingomyelinase with accumulation of sphingomyelin in the lysosomes** of histiocytes in the brain, bone marrow, spleen, and liver.

NIEMANN–PICK DISEASE

ID/CC	A 2-year-old female is referred to a pediatric clinic for evaluation of **lethargy, weakness, and persistent anemia** that has been unresponsive to treatment with iron, folic acid, and vitamins C and B_{12}.
HPI	She is the third-born child of a healthy white couple; her mother had an uneventful pregnancy and a eutopic delivery. Both brothers are healthy.
PE	**Low weight and height** for age; **marked pallor**; flaccidity and lethargy; **sleepiness**. No focal neurologic signs; lungs clear; heart sound with slight aortic systolic ejection murmur (due to anemia); abdomen soft; no masses; no hepatomegaly; spleen barely palpable; no lymphadenopathy.
Labs	CBC: **megaloblastic anemia**; elevated mean corpuscular volume. UA: increased orotic acid excretion with formation of **orotic acid crystals**.
Treatment	Administration of **uridine** and cytidine. Steroids.
Discussion	Orotic aciduria is an **autosomal-recessive disorder of pyrimidine synthesis**; it is caused by a deficiency of the enzyme system orotidylic pyrophosphorylase-orotidylic decarboxylase with resultant megaloblastic anemia due to impaired synthesis of nucleic acids necessary for hematopoiesis.

ID/CC A 3-year-old male presents with progressive **mental retardation, vomiting**, and **hyperactivity** with purposeless movements.

HPI The child developed normally for the first 2 to 3 months. He is fairer than his siblings and, unlike them, has blue eyes. He was born outside the United States and did not undergo any screening for congenital disorders.

PE Child is **blond with blue eyes, perspires** heavily, is mentally retarded, and has **peculiar "mousy" odor; hypertonia** with hyperactive deep tendon reflexes on neurologic exam.

Labs Guthrie test (bacterial inhibition assay method) positive (due to **increased blood phenylalanine levels**); increased urinary phenylpyruvic and ortho-hydroxyphenylacetic acid; **normal concentration of tetrahydrobiopterin.**

Imaging XR: delayed bone age.

Treatment Diet formulas low in phenylalanine. Avoid aspartame. **Tyrosine supplementation**.

Discussion PKU is an **autosomal-recessive disorder** caused by a **deficiency of the enzyme phenylalanine hydroxylase**. A neonatal screening program for the detection of PKU is in effect throughout the United States.

GENETICS

PHENYLKETONURIA (PKU)

ID/CC   A 7-year-old female presents with anxiety, **dizziness, sweating**, and nausea **following** brief episodes of **exercise**.

HPI   These symptoms are **relieved by eating** and do not occur if the patient is frequently fed small meals.

PE   Physical exam unremarkable.

Labs   **Hypoglycemia** following brief fasting; alanine fails to increase blood sugar; fructose or glycerol administration restores blood glucose to normal.

Micro Pathology   Liver biopsy for enzyme assays reveals deficiency of phosphoenolpyruvate carboxykinase, an enzyme of gluconeogenesis; **no excess glycogen storage** revealed.

Treatment   Frequent small meals to prevent episodes of hypoglycemia.

Discussion   Phosphoenolpyruvate carboxykinase (PEPCK) deficiency prevents pyruvate from being converted to phosphoenolpyruvate. This deficiency interferes with gluconeogenesis from 3-carbon precursors (e.g., alanine) that enter the gluconeogenetic pathway at or below the pyruvate level.

ID/CC A **2-month-old** child is brought to the pediatrician because of **failure to gain weight**, increasing **weakness**, insufficient strength to breast feed, and **lethargy**.

HPI He is the second-born son of a healthy white couple; his mother's pregnancy and delivery were uneventful.

PE Mild cyanosis; shallow respirations; increase in size of tongue (MACROGLOSSIA); moderate hepatomegaly; **significant generalized muscular flaccidity**.

Labs CBC: normal. Lytes: normal. **Glucose**, BUN, creatinine **normal**. ECG: **short P-R; wide QRS; left-axis deviation**.

Imaging CXR: **extreme cardiomegaly** and congestive heart failure.

Gross Pathology Significant increase in size and weight of heart (up to five times normal); to lesser extent, hepatomegaly.

Micro Pathology Extensive intracytoplasmic and lysosomal deposition of glycogen on myocardial fibers as well as in striated muscle fibers, kidney, and liver.

Treatment Poor prognosis; associated with early death from **cardiopulmonary failure**.

Discussion Pompe's disease is a **type II glycogen storage disease** (generalized). This fatal disorder is caused by an **autosomal-recessive deficiency in the lysosomal** enzyme (only glycogenolysis with lysosomal involvement) **α-1,4-glucosidase** (ACID MALTASE), with resulting accumulation of structurally normal glycogen in the heart, muscle, kidney, and liver.

POMPE'S DISEASE

ID/CC	A 40-year-old **male** visits his family doctor because of a **chronic, recurrent rash** on his hands, face, and other **sun-exposed areas**; the patient's **urine turns dark brown-black if left standing**, and he has noticed that recurrences coincide with **alcohol intake**.
HPI	He reports having used **hexachlorobenzene as a pesticide** for some years (a fungicide shown to be associated with porphyria cutanea tarda).
PE	Skin erythema with **vesicles and bullae on sun-exposed areas**; skin at these sites is friable and shows presence of whitish plaques ("MILIA") (due to photosensitizing effect of uroporphyrin); skin of face also shows hypertrichosis and hyperpigmentation.
Labs	Watson–Schwartz test negative. UA: **markedly elevated urinary uroporphyrin levels; slightly elevated urinary coproporphyrin levels**. Fecal isocoproporphyrin normal; **elevated transferrin, serum and hepatic iron**; elevated serum transaminases.
Gross Pathology	Liver shows siderosis, bullae, fibrosis, and inflammatory changes.
Micro Pathology	Skin biopsy demonstrates iron deposits, intense porphyrin fluorescence, and long, thin cytoplasmic inclusions.
Treatment	Repeated phlebotomies; avoidance of sunlight, alcohol, iron, and estrogens. Low dose antimalarials.
Discussion	Porphyria cutanea tarda (PCT), in contrast to other hepatic porphyrias, is more common among men than women. PCT is caused by partial loss of activity of **hepatic uroporphyrinogen decarboxylase**; lesions are caused by overproduction and excretion of uroporphyrin.

ID/CC A 5-year-old girl is referred to a hematologist for an evaluation of chronic **anemia that has been unresponsive to nutritional supplementation**.

HPI Both parents are clinically normal and are **first cousins** who are **Amish**. The patient has no history of passage of dark-colored urine or recurrent infections.

PE Low weight and height for age; pallor; mild **jaundice**; spleen barely palpable; liver not enlarged.

Labs CBC/PBS: **anemia; markedly increased reticulocyte count**; peripheral blood reveals macro-ovalocytosis with a few echinocytes; no sickle cells or spherocytes seen. Hyper-bilirubinemia (primarily unconjugated). UA: urinary hemosiderin present. Reduced serum haptoglobin; **diminished activity of pyruvate kinase** in RBCs on spectrophotometry.

Treatment **Exchange transfusions**. Splenectomy.

Discussion Pyruvate kinase deficiency is inherited as an **autosomal-recessive** trait and usually produces mild symptoms (hemolytic anemia); **2,3-diphosphoglycerate accumulates**, shifting the hemoglobin-oxygen dissociation curve to the right (due to reduced affinity of RBCs for oxygen).

PYRUVATE KINASE DEFICIENCY

ID/CC	A **6-month-old** male is brought to a pediatrician for evaluation of **listlessness**, lethargy, and **fixed gaze**.
HPI	His parents are **Ashkenazi Jews**.
PE	**Excessive extensor startle response to noise** (HYPERACUSIS); child is sleepy and hypotonic with poor head control and a fixed gaze; appears to have translucent skin; **cherry-red macular spot** found on funduscopic exam.
Gross Pathology	Diffuse gliosis; cerebral and macular degeneration; up to 50% increase in brain weight (due to deposition of sphingolipid).
Micro Pathology	Neuronal swelling with cytoplasmic **deposits of gangliosides** (ZEBRA BODIES).
Treatment	Poor prognosis; patients usually die of pneumonia before reaching the age of 3.
Discussion	Tay–Sachs disease is an autosomal-recessive disorder of sphingolipid metabolism characterized by the **absence of the enzyme hexosaminidase A**, producing excessive storage of ganglioside GM-2 in lysosomes **restricted to the cells of the central nervous system. Ganglioside GM-2** is a glycosphingolipid with sphingosine, a long-chain basic molecule, as its backbone along with an attached sugar and a terminal N-acetylglucosamine. Prenatal diagnosis can be made at the 14th week of pregnancy.

ID/CC A 7-year-old male is brought to a pediatrician for evaluation of episodes of **fatigue**, restlessness, anxiety, nausea, **lightheadedness**, vomiting, and **sweating**.

HPI The symptoms appear when he does not eat frequent meals and subside while he is eating. He also has a history of **bruising easily**.

PE Patient has **"doll-face" facies**; weight low for age; tendon xanthomas; purpuric patches over skin; **marked hepatomegaly**.

Labs Lactic acidosis; hyperlipidemia; **marked increase in serum uric acid** (patient may exhibit gout symptoms); **marked hypoglycemia**; prolonged bleeding time; **lack of rise in serum glucose following SC epinephrine** or IV glucagon but striking increase in lactate; normal urinary catecholamines.

Imaging US: hepatomegaly; kidneys also enlarged bilaterally.

Gross Pathology **Liver and kidneys enlarged** (vs. type III glycogen storage disease, or Cori's disease, in which there is no renal involvement).

Micro Pathology Hepatocytes containing variable-sized **glycogen-lipid droplets** on liver biopsy; nuclear glycogenosis seen; large glycogen deposits in kidney; skeletal and cardiac muscle not involved (vs. type V glycogen storage disease, or McArdle's disease, in which skeletal muscle is involved).

Treatment Frequent meals to prevent hypoglycemia.

Discussion Von Gierke's disease is an autosomal-recessive glycogen storage disease (type I) resulting from a **deficiency of glucose-6-phosphatase** and accumulation of structurally normal glycogen in the liver and kidneys.

VON GIERKE'S DISEASE

ID/CC	A 10-year-old girl is brought by her parents to a dermatologist because of a recent **change in color and increase in size of a warty lesion** on her face.
HPI	She has been suffering from **excessive sensitivity to sunlight** and thus does her best to avoid the sun as much as possible.
PE	Abundant **freckles** on all sun-exposed areas; **telangiectases**; areas of redness (ERYTHEMA) and hypopigmentation; **hyperkeratosis** on face and dorsum of hands; hard, nodular lesion on right cheek; no regional lymphadenopathy.
Labs	Basic lab work normal.
Gross Pathology	Generalized hyperpigmentation with desquamative spots on sun-exposed areas.
Micro Pathology	Biopsy of cheek lesion reveals hyperkeratosis with melanin deposition; **squamous cell carcinoma**.
Treatment	Avoidance of sunlight, protection against sunlight. Surgical removal of cancer.
Discussion	Xeroderma pigmentosum is an **autosomal-recessive** disorder that is usually manifested in childhood. It is characterized by excessive sensitivity to ultraviolet light due to **impaired endonuclease excision repair mechanism** of **ultraviolet light-damaged DNA** UV light causes (cross linking of pyrimidine residues) in dermal fibroblasts. There is a marked tendency to develop **skin cancer** (squamous cell and basal cell carcinoma).

ID/CC A 40-year-old woman presents with **weakness, easy fatigability, nausea**, and **diarrhea**.

HPI She has had a long and severe course of rheumatoid arthritis for which she has been taking **methotrexate** (a folic acid antagonist).

PE VS: normal. PE: **pallor**; mild tongue inflammation (GLOSSITIS); funduscopic exam normal; chest sounds within normal limits; abdomen shows no hepatosplenomegaly; no lymphadenopathy; **no neurologic signs** (vs. vitamin B_{12} deficiency megaloblastic anemia).

Labs CBC: **hypersegmented PMNs** (> 5 to 7 lobes); **megaloblastic RBCs** (mean corpuscular volume >100); **vitamin B_{12} level normal; folate** level **in RBCs low** (vs. vitamin B_{12} deficiency megaloblastic anemia).

Treatment Folic acid supplementation.

Discussion Folic acid is found mainly in green leaves and is important for the **synthesis of DNA and RNA**. It also acts as a coenzyme for 1-carbon transfer and is involved in **methylation reactions**. Deficiency is associated with alcoholism, pregnancy (MEGA-LOBLASTIC ANEMIA OF PREGNANCY), dietary deficiencies, and drugs such as TMP-SMX, methotrexate, phenytoin, and proguanil.

ANEMIA—FOLATE DEFICIENCY

ID/CC A 1-year-old infant presents at a clinic with **lassitude, poor muscle tone**, and delayed motor development.

HPI The mother is a known IV drug user and has two older children who are in the custody of the state social services agency.

PE VS: tachycardia; tachypnea. PE: **pallor**; partial alopecia; ulceration of skin at corners of mouth (CHEILOSIS); smooth tongue; **nails break easily** and **are spoon shaped** (KOILONYCHIA).

Labs PBS: abnormally **small and pale RBCs** (MICROCYTIC, HYPOCHROMIC ANEMIA); RBCs of different sizes (ANISOCYTOSIS) and different shapes (POIKILOCYTOSIS). **Decreased serum iron**; increased total iron-binding capacity and reduced percentage saturation; artificial **increased transferrin** (due to attempt to transport as much iron as possible); **low serum ferritin** (due to insufficient iron stored as ferritin).

Gross Pathology Atrophic glossitis.

Micro Pathology Erythroid hyperplasia with **decreased bone marrow iron stores on Prussian blue staining** (vs. anemia of chronic disease, which is characterized by increased iron stores).

Treatment Control cause of iron deficiency; supplemental iron.

Discussion Iron-deficiency anemia is the most common cause of **chronic blood loss**, usually gastrointestinal or gynecologic; it is secondary to a deficiency of iron required for normal hemoglobin synthesis. Differentiate from anemia of chronic disease, in which ferritin is high and transferrin is low.

ID/CC	A 58-year-old black female complains of **weakness, dizziness**, anorexia, nausea, and occasional vomiting over the past 3 months.
HPI	She has also experienced **shortness of breath** (due to diminished oxygen-carrying capacity) as well as **numbness and tingling** in the extremities (due to megaloblastic peripheral neuritis).
PE	Slightly icteric eyes; hepatosplenomegaly; smooth, beefy-red tongue (GLOSSITIS); **loss of balance, vibratory, and position sense** in both lower extremities (due to posterior and lateral column involvement; vs. folic acid deficiency).
Labs	CBC: **macrocytic**, hypochromic anemia (MCV > 100); **leukopenia** (4,000) with **hypersegmented neutrophils**; thrombocytopenia. Hyperbilirubinemia (2.5 mg/dL; normal 0.1 to 1.0 mg/dL); **achlorhydria** (no hydrochloric acid in gastric juice); positive Schilling test; **low** blood **vitamin B_{12} levels**; RBC folate normal.
Gross Pathology	Dorsal and lateral spinal columns are small, with axonal degeneration; flat, atrophic mucosa; loss of rugal folds in stomach; increased red marrow in bone.
Micro Pathology	**Megaloblastic** and **hypercellular bone marrow** with erythroid hyperplasia; accumulation of hemosiderin in Kupffer cells; chronic atrophic gastritis.
Treatment	Parenteral vitamin B_{12}.
Discussion	Pernicious anemia is megaloblastic anemia caused by malabsorption of vitamin B_{12} because of **lack of intrinsic factor** in gastric juice (intrinsic factor, secreted by parietal cells, is indispensable for vitamin B_{12} absorption). **Antibodies against gastric parietal cells** are almost invariably present in the adult form of the disease.
Atlas Links	[UCV1] **H-BC-077** [UCV2] IM1-043

ANEMIA—VITAMIN B_{12} DEFICIENCY

ID/CC   A 31-year-old **black male** who works as a Peace Corps volunteer in Ghana visits his medical officer complaining of extreme **weakness and fatigue**; he also complains of a **yellowing of his skin and slight fever**.

HPI   He was prescribed **primaquine** for radical treatment of malaria (due to *Plasmodium vivax*).

PE   VS: tachycardia (HR 95). PE: mild jaundice; circumoral and nail bed **pallor**; no hepatosplenomegaly; remainder of PE normal.

Labs   **Elevated indirect bilirubin**. CBC/PBS: **low hemoglobin and hematocrit** (9.3/33) with reticulocytosis (HEMOLYSIS); **spherocytes** in peripheral blood smear; **Heinz bodies** (precipitated hemoglobin) in RBCs; **low blood levels of G6PD** (diagnostic). UA: hemoglobinuria.

Treatment   Withdrawal of offending drug.

Discussion   Glucose-6-phosphate dehydrogenase (G6PD) deficiency is an **X-linked recessive** disorder seen in about 15% of American black males. With infections or exposure to certain drugs (e.g., sulfa drugs, antimalarials, nitrofurantoin), patients deficient in G6PD present with a **hemolytic anemia due to increased RBC sensitivity to oxidant damage**. G6PD is the rate-limiting enzyme in the HMP shunt that provides NADPH.

ID/CC	A 7-year-old male is brought to the emergency room because of weakness and the **spontaneous appearance** of painful swelling of both knee joints (due to hemarthrosis) as well as black, tarry stools (GI bleeding).
HPI	The child has a **history of prolonged bleeding following minor injuries**. His maternal uncle died of a "bleeding disorder."
PE	Pallor; swollen, erythematous, tender knee joints with blood accumulation in synovial capsule (HEMARTHROSIS); numerous **bruises** seen at areas of minimal repeated trauma.
Labs	Bleeding time and PT normal; **prolonged PTT**; reduced levels of factor VIII on immunoassay; synovial fluid hemorrhagic.
Imaging	XR: bilateral knee effusions.
Micro Pathology	Synovium may show hyperplasia with hemosiderin in synovial macrophages.
Treatment	Nonpharmacologic therapy involves patient education, **avoidance of contact sports, avoidance of aspirin and other NSAIDs** (due to antiplatelet aggregating effect), **orthopedic evaluation** and physical therapy, and **hepatitis vaccination. Factor VIII supplementation** is effective in controlling spontaneous and traumatic hemorrhage. **Desmopressin** may be used prophylactically in patients with mild hemophilia, prior to minor surgical procedures. **Aminocaproic acid** may be used to stop bleeding that is unresponsive to factor VIII or desmopressin.
Discussion	Hemophilia A is an **X-linked recessive disorder** that is manifested by bleeding and is due to a **deficiency in coagulation factor VIII**.
Atlas Link	UCV2 PED-018

HEMOPHILIA, TYPE A

ID/CC   A 9-month-old infant is brought to the pediatrician because of **jaundice**, lethargy, and **easy fatigability**.

HPI   The parents of the child are immigrants of **northern European origin**.

PE   Pallor; mild jaundice; palpable **splenomegaly**.

Labs   CBC/PBS: microcytic **anemia**; small, **rounded**, dark RBCs lacking central pallor; negative Coombs' test. **Elevated indirect bilirubin; increased reticulocytes; increased mean corpuscular hemoglobin count** (> 35); decreased MCV; abnormal RBC osmotic lysis test.

Treatment   **Folic acid. Splenectomy**.

Discussion   Hereditary spherocytosis is a congenital, **autosomal-dominant** disorder characterized by **hemolytic anemia with spherical RBCs** and splenomegaly. It is caused by a **defect in RBC membrane spectrin** with loss of the normal biconcavity and a higher rate of splenic sequestration and hemolysis. If left untreated, it may give rise to pigment **gallstones** and **cholecystitis**.

ID/CC	A newborn male in the normal nursery is noted to be **cyanotic**; the pediatrician is called even though the child does not seem to be in respiratory distress.
HPI	That morning he had undergone circumcision (a **benzocaine** ointment was used).
PE	**Cyanotic**; lungs clear and well ventilated; heart sounds rhythmic; no murmurs heard; no cardiopulmonary problems evident.
Labs	CBC/Lytes: normal. Platelets, LFTs, BUN, and creatinine normal. ABGs: **PO$_2$ normal. Methemoglobin level 18% total hemoglobin**.
Imaging	CXR: normal
Treatment	Oxygen for acute symptoms. **Methylene blue** (increases activity of methemoglobin reductase).
Discussion	Methemoglobin is an oxidized (FERRIC) form of hemoglobin that cannot function properly as a carrier of oxygen. The diminished oxygen-carrying capacity that results produces headache, lightheadedness, and dyspnea. Drugs such as dapsone and benzocaine as well as dyes such as anilines oxidize hemoglobin to its ferric form, as do **deficiencies of NADH methemoglobin reductase**. In neonates there is a transient deficiency of this enzyme, and HbF is more susceptible than HbA to oxidation.

METHEMOGLOBINEMIA

ID/CC A 19-year-old **male** comes to see the nurse at the college health department complaining of **abdominal and lumbar pain**, which characteristically occurs **when he takes his multivitamin pills** two times a week (iron, infections, and vaccination are precipitating factors); he has also noticed **dark brown urine the morning** after he has the pain (due to hemolysis).

HPI He has just left his parents to go to college and is excited about his newfound freedom; he likes to drink excessive amounts of beer.

PE Marked **pallor**; lung fields clear to auscultation; heart sounds normal; abdomen soft and nontender with no masses or peritoneal signs; no focal neurologic signs.

Labs CBC: normocytic, normochromic **hemolytic anemia** with reticulocytosis. **Hemoglobinemia and hemoglobinuria; sucrose hemolysis test positive; acidified serum test positive** (HAM'S TEST); **decreased haptoglobin; elevated LDH; decreased leukocyte alkaline phosphatase.**

Gross Pathology Hemosiderosis of liver, spleen, and kidney.

Treatment Steroids, transfusion of saline-washed RBCs during crises. Oral iron supplementation may be useful but should be used cautiously, as it may precipitate transient hemolysis. Similarly, heparin may accelerate hemolysis, but its use in thrombotic complications appears warranted.

Discussion Paroxysmal nocturnal hemoglobinuria is an **acquired** defect of the red blood cell membrane, making erythrocytes unusually sensitive to serum complement (there is also increased binding of C3b). It is characterized by episodes of hemolysis with hemoglobinuria that occur during sleep because of carbon dioxide retention (which lowers the pH, thus enhancing complement activity; first voided urine in the morning is red-brown). Patients are also predisposed to developing venous thromboses.

PAROXYSMAL NOCTURNAL HEMOGLOBINURIA

ID/CC	A 33-year-old woman presents to a clinic with **marked weakness** (due to hypokalemia).
HPI	Two years ago, she underwent a ureterolithotomy for **renoureteral stones**.
PE	VS: tachypnea. PE: **generalized muscle weakness**; heart sounds with a few skipped beats (hypokalemia gives rise to severe arrhythmias); diminished intestinal peristalsis; no peritoneal signs.
Labs	Lytes: increased urinary potassium excretion (due to insufficient hydrogen ion available, with potassium exchanged for sodium), resulting in **marked hypokalemia** (2.3 mEq/L). ABGs: decreased HCO_3 (due to failure to maintain normal gradient of hydrogen ions in distal renal tubules, with HCO_3 loss); **hyperchloremic metabolic acidosis** (normal anion gap). Normal serum calcium; high alkaline phosphatase. UA: urine alkaline; hypercalciuria.
Imaging	KUB: radiopaque left kidney stones; medullary renal calcification.
Gross Pathology	Nephrocalcinosis.
Treatment	Bicarbonate; potassium and vitamin D.
Discussion	Metabolic acidosis is caused by renal tubular defects in transport. Type I (distal) involves selective deficiency of tubular H^+ secretion (produces typical hyperchloremic-hypokalemic acidosis with normal anion gap). Type II (proximal) involves the inability to reabsorb HCO_3 (also hypokalemic). Type III entails the inability to produce NH_3 due to persistently low GFR volumes (normokalemic). Type IV is due to primary or drug-induced hypoaldosteronism (hyperkalemic).

RENAL TUBULAR ACIDOSIS

ID/CC A 38-year-old electrician is rushed to the emergency room after receiving an accidental high-voltage **electric shock** while fixing a power line.

HPI On admission, a Foley catheter is inserted, yielding **reddish-brown urine** (due to myoglobin).

PE VS: tachycardia; BP normal. PE: confusion; disorientation; patient complains of **muscle pain** in right arm, leg, and buttock; hand severely swollen and has an oblique-shaped **burn**; "outlet wound" located in right gluteal region and ankle.

Labs **Markedly increased serum BUN and creatinine** (due to acute tubular necrosis); urea normal. Lytes: **hyperkalemia**. Hyperphosphatemia; hyperuricemia; hypocalcemia (due to calcium binding to necrotic muscle); **increased serum CK** (due to muscle destruction); **myoglobinuria**.

Treatment Urine alkalinization (with IV bicarbonate); vigorous **rehydration** (to prevent pigment deposition and acute tubular necrosis); mannitol; prevent further muscle damage from compartment syndromes (evaluate need for fasciotomy). Correct electrolyte abnormalities. Hemodialysis may be required in severe cases.

Discussion Myoglobinuria and reduced renal perfusion from volume depletion may cause acute tubular damage. Other causes of rhabdomyolysis (**destruction of striated muscle**) include crush injuries, heroin overdose, prolonged unconsciousness in one position, arterial occlusion, alcohol abuse, and seizures.

RHABDOMYOLYSIS

ID/CC	A 4-year-old **male** is brought to the pediatric clinic because of **easy fatigability and difficulty walking** of a few months' duration.
HPI	The child's mother has noticed that his calves have increased in size **(pseudohypertrophy)**.
PE	Child well developed but shows **proximal muscle weakness** in shoulder and pelvic girdle; difficulty standing and walking; "climbs up on himself" to rise from sitting to standing (GOWERS' SIGN).
Labs	**CK, LDH, and glucose phosphoisomerase elevated; absent dystrophin** expression on immunostain of **muscle biopsy**.
Gross Pathology	Replacement of normal muscle protein with fibrofatty tissue, giving rise to pseudohypertrophy.
Micro Pathology	Degeneration and atrophy of muscle fibers with ringed fibers surrounding normal tissue.
Treatment	Prognosis is poor, with disability occurring within a few years and death by the early 20s. Treatment is supportive. Refer for genetic counseling.
Discussion	Duchenne's muscular dystrophy is an **X-linked recessive** disorder characterized by a deficiency in muscle **dystrophin**, a subsarcolemmal cytoskeletal protein that stabilizes the sarcolemma during contraction and relaxation. Its course is relentlessly progressive, ending in death from cardiac and respiratory muscle involvement.
Atlas Links	UCV1 **PG-BC-085, PM-BC-085** UCV2 PED-049

DUCHENNE'S MUSCULAR DYSTROPHY

ID/CC	A 40-year-old male is brought to the ER from a **bar** because of **confusion** after falling from a bar stool.
HPI	The patient's friends say that his diet consists mainly of **alcoholic drinks**. They also state that he has told **detailed and believable stories about his past adventures** that have **subsequently been found to be untrue** (CONFABULATION). His **short- and long-term memory** is **severely impaired**.
PE	**Ataxia; oculomotor abnormalities**, including **nystagmus** and **ophthalmoplegia**.
Labs	CBC: macrocytic anemia (most likely secondary to folate deficiency). **Low thiamine (B_1) levels**.
Gross Pathology	**Bilateral atrophy of mammillary bodies** and thalamus.
Micro Pathology	Neuronal degeneration in mammillary bodies and thalamus.
Treatment	Immediate thiamine (B_1) administration parenterally; the sooner treatment is administered, the less permanent the sequelae. Alcoholics should also receive oral or IV folate as well as a multivitamin. Monitor carefully for **delirium tremens** secondary to alcohol withdrawal.
Discussion	Wernicke's encephalopathy is for the most part **reversible with thiamine treatment**. A delay in treatment may cause progression to Korsakoff's psychosis with permanent dementia. Patients rarely return to normal. Patients also often have **wet beriberi** (high-output cardiac failure), **dry beriberi** (peripheral neuropathy with impairment of distal motor and sensory function), and cerebral beriberi (motor and cognitive impairment). **Wernicke's encephalopathy** consists of a triad of **confusion, ataxia, and ophthalmoplegia**. Korsakoff's psychosis is characterized by **retrograde/anterograde amnesia and confabulation**.
Atlas Link	UCVI **PG-BC-087**

ID/CC	After a routine pelvic exam, a 23-year-old **female** is referred by her family physician to an endocrinologist for an evaluation of **"lack of a palpable cervix."**
HPI	The patient states that she **has never had a menstrual period.**
PE	**Bilateral breast tissue present; absence of pubic and axillary hair; vagina ends in blind pouch**; clitoromegaly; small atrophic **testis found** on right inguinal canal.
Labs	**Increased LH** and **testosterone.** Karyotype: **46,XY.**
Imaging	US: uterus and ovaries absent.
Treatment	**Treat as woman**, resect cryptorchid testis and look for the intra-abdominal one (due to high risk of maliznancy).
Discussion	Also known as androgen insensitivity syndrome, testicular feminization is characterized by a genotypically male individual (KARYOTYPE 46,XY) who presents with a female body habitus with breast development and cryptorchidism; it is due to a Y chromosome gene defect that causes the **testosterone receptor protein to be unresponsive to androgenic stimulation.**

PRIMARY AMENORRHEA—TESTICULAR FEMINIZATION

ID/CC   A 20-year-old black female visits her gynecologist because she thinks she might be pregnant because of **lack of menses for the past 4 months**.

HPI   She is a **pentathlon athlete** who is training to compete in her home state's tournament next fall. She is sexually active, uses the "rhythm method" for birth control, **and has never missed a menstrual period**.

PE   No breast enlargement; no softening of cervix; no bluish discoloration of cervix (both presumptive signs of pregnancy); no abdominal or pelvic masses or palpable uterus; no hirsutism or virilization.

Labs   Serum and urinary β-hCG **negative for pregnancy**; serum prolactin and TSH normal; **decreased serum FSH**; no withdrawal bleeding after administration of progesterone.

Imaging   XR, skull: normal sella.

Treatment   Advise patient to either gain enough weight to restore menses or take oral contraceptives to prevent osteoporosis.

Discussion   The **most common cause of secondary amenorrhea is pregnancy**. Women who are involved in vigorous physical exercise and who lose weight may present with a functional gonadotropin deficit. When body weight falls > 15% of ideal weight, GnRH secretion from the hypothalamus is decreased, producing a secondary amenorrhea. The inhibitory effect of estrogens on bone resorption is also lost, predisposing patients to an increased risk for osteoporosis.

ID/CC	A 5-year-old white male is brought to the emergency room with a fracture of his right forearm that he sustained after falling off a couch.
HPI	This is the **fifth bone fracture** that the child has sustained **in the past 2 years**.
PE	**Bluish sclera**; right leg and right arm slightly deformed from poor healing of past fractures; mild **kyphosis and scoliosis** of thoracic spine; **hypotonia and laxity** of right leg and arm; **partial conduction deafness** in both ears.
Imaging	XR: fracture of radius and ulna with evidence of osteopenia.
Micro Pathology	**Marked thinning of bone cortices** (EGGSHELL CORTEX) and rarefaction of trabeculae (due to **abnormal synthesis of type I collagen**); abnormal softening of tooth enamel.
Treatment	Supportive.
Discussion	Also called brittle bone disease, osteogenesis imperfecta is an **autosomal-dominant** disorder of type I collagen synthesis in which there is deficient ossification due to inadequate osteoid formation.
Atlas Link	UCV2 PED-052

OSTEOGENESIS IMPERFECTA

ID/CC A 45-year-old male chess player is brought to the emergency room complaining of acute **nausea**; he has **vomited** five times, feels very **lightheaded**, and has a severe **headache**.

HPI He went out **drinking** last night to celebrate his victory in a chess tournament he attended last week in Mexico. While in Mexico, he contracted acute amebiasis that is currently being treated with **metronidazole**.

PE VS: marked tachycardia (HR 120); **hypotension** (BP 90/60). PE: anxious, dehydrated, and confused with severe nausea.

Labs CBC/LFTs: normal. Amylase normal. ABG/Lytes: mild hypokalemia and metabolic alkalosis (due to vomiting).

Treatment Supportive, IV fluids, antiemetics, discontinuance of alcohol.

Discussion Ethanol is degraded by alcohol dehydrogenase to acetaldehyde, which in turn is degraded to acetic acid by another acetaldehyde dehydrogenase. This **acetaldehyde dehydrogenase is inhibited** by disulfiram, resulting in the **accumulation of acetaldehyde**, which produces nausea, vomiting, headache, and hypotension (ANTABUSE EFFECT). Metronidazole, some cephalosporins, and other drugs have an Antabuse-like effect when consumed concomitantly with alcohol.

ID/CC	A **premature** (32-week-old) white male infant is brought to the intensive care unit after a **cesarean** delivery.
HPI	His mother had third-trimester **bleeding** and contractions that did not stop with rest and conservative treatment.
PE	VS: tachypnea. PE: child weighs 3.8 lb; **cyanosis; dyspnea**; uses accessory muscles of respiration; **nasal flaring**.
Labs	ABGs: hypoxemia; hypercapnia. Decreased lecithin/sphingomyelin (L/S) ratio (L/S ratio normally > 2; 1.5 to 2.0 in 40% of newborns with respiratory distress syndrome).
Imaging	CXR: bilateral **reticular pulmonary infiltrates** and atelectasis.
Gross Pathology	Generalized atelectasis in purple-colored lung; eosinophilic fibrinous hyaline membrane formation.
Treatment	Ventilatory support, fluid, acid-base and electrolyte balance, antibiotics; administration of surfactant; steroids before birth to speed lung maturity.
Discussion	**Respiratory distress syndrome** of the newborn is the most common cause of death in premature infants. It is due to a **deficiency of surfactant**, a lipoprotein produced by type II pneumocyte cells that contains the phospholipid dipalmitoyl lecithin. Fetal lung maturity may be measured by the L/S ratio. The syndrome might be prevented by giving **betamethasone** to pregnant women, since type II pneumocyte cell differentiation is dependent on steroids. Complications include patent ductus arteriosus, pulmonary air leaks, and bronchopulmonary dysplasia.
Atlas Link	UCV1 PG-BC-083

HYALINE MEMBRANE DISEASE

ID/CC A 36-year-old female **nonsmoker** visits her family doctor because she has become increasingly **short of breath** (DYSPNEA); her symptoms first appeared only during exercise but now occur even when she is at rest.

HPI She also complains of frequent URIs and moderate **weight loss**.

PE **Thin** female with **increased anteroposterior diameter of chest** (BARREL-SHAPED CHEST); decreased breath sounds bilaterally; **hyperresonance** to percussion; retardation of expiratory flow.

Labs CBC: increased hematocrit. PFTs: $FEV_1/FVC < 75\%$ (diagnostic of airflow obstruction). ECG: right ventricular hypertrophy.

Imaging CXR: hyperlucent lung fields; flattening of diaphragm and decreased lung markings at periphery.

Gross Pathology Destruction of alveolar walls distal to the terminal bronchiole with hyperaeration (EMPHYSEMA); **panacinar type** (COTTON CANDY LUNG); more severe at lung bases.

Treatment Standard treatment for COPD patients. Replacement therapy with α_1-protease inhibitor.

Discussion Pollutants, cigarette smoke, and infections increase PMNs and macrophages in the lung and thus produce a number of proteolytic enzymes. Damage to lung tissue due to these enzymes is controlled by the globulin α_1-**antitrypsin**, which inhibits trypsin, neutrophil, elastase, and collagenase. A deficiency of this enzyme causes excessive lung tissue destruction and **panacinar emphysema** (cigarette smoking is associated with the centrilobular type). Patients may also develop **liver damage**.

ALPHA-1-ANTITRYPSIN DEFICIENCY

ID/CC	A 23-year-old female college student is brought to the emergency room because of **numbness** of her face and feet together with a **sensation of suffocation** and **stiff twisting of the hands** (CARPOPEDAL SPASM); these symptoms arose following an argument with her boyfriend.
HPI	A friend reports that the patient has a **history of anxiety-induced colitis, gastritis, and migraine**.
PE	VS: **marked tachypnea** (RR 40); tachycardia (HR 90); hypertension (BP 140/90). PE: patient **apprehensive** and anxious; physical exam otherwise normal.
Labs	ABGs: **low P_{CO_2}; respiratory alkalosis** (cause of tetany); **low bicarbonate** (to compensate for primary lowering of P_{CO_2}).
Imaging	CXR: normal
Treatment	Have patient **breathe in and out of a bag** or give 5% CO_2 mixture.
Discussion	Anxiety hyperventilation is a common occurrence in ERs. The anxiety state produces an increase in the frequency of respirations (HYPERVENTILATION), causing a lowering of P_{CO_2}; the resulting respiratory alkalosis produces an unstable depolarization of the distal segments of motor nerves with symptomatic tetany. Alkalosis also sets in motion a compensatory decrease in bicarbonate level to maintain pH as close to normal as possible.

HYPERVENTILATION

ID/CC A 36-year-old divorcee living in rural Maine is brought by ambulance to the ER with her two children, who were **all found unconscious in her home** by military personnel.

HPI A recent "El Niño" produced bad weather that resulted in a power failure; as a result, she had been using charcoal and a **wooden stove inside her house** for heating purposes.

PE **Skin bright red** (CHERRY-RED CYANOSIS); pulse arrhythmic; patient regains consciousness soon after administration of 100% oxygen but remains drowsy, **disoriented**, and nauseous and complains of a severe **headache** (due to cerebral edema); **hyperreflexia** noted as well as positive Romberg's test.

Labs **Increased carboxyhemoglobin** ($> 25\%$). ABGs: metabolic acidosis.

Imaging CT/MR: bilateral globus pallidus lesions.

Treatment One hundred percent **oxygen**, assisted ventilation if necessary. Hyperbaric oxygen chamber.

Discussion Common sources of CO are car exhaust, pipes, and fires. Carbon monoxide has a much greater affinity for hemoglobin than oxygen (250 times more). If patient is pregnant, damage to the fetus is devastating (HbF has greater affinity to CO than HbA). Long-term side effects such as memory problems, lack of coordination, and even convulsions are common after intoxication.

CARBON MONOXIDE POISONING

ID/CC	A 52-year-old obese white male comes to his family doctor complaining of severe pain in the **first metatarsophalangeal (MTP) joint** (PODAGRA) that began at **night** after an episode of binge **eating** and **drinking**.
HPI	He admits to being an avid **meat** eater and drinks **red wine** every night. His history is significant for removal of **kidney stones** (uric acid stones).
PE	VS: **fever** (38.2°C). PE: **right MTP joint** red, hot, and swollen; painful to active and passive motion; **tophaceous** deposits in left ear and olecranon bursitis.
Labs	**Elevated serum uric acid**. UA: urate crystals. **Increased ESR**. CBC: leukocytosis with neutrophilia.
Imaging	XR: punched-out erosions in right big toe at MTP joint, producing **"overhanging"** spicules.
Gross Pathology	Tophi are white, soft, nodular masses of urate deposits with calcifications seen mainly in synovial membranes, tendon sheaths, and ear cartilages.
Micro Pathology	Tophi and synovial fluid aspiration show characteristic negatively **birefringent, needle-shaped crystals** of uric acid salts; giant cell formation with neutrophilic infiltration.
Treatment	Colchicine for acute stage; administer hourly until negatively diarrhea occurs or pain disappears; combine with aspirin or NSAIDs. Long-term treatment with allopurinol and/or probenecid.
Discussion	Gout is a disorder of purine metabolism with a resulting increase in serum uric acid level and deposits in several tissues; 10% to 20% of cases may develop nephrolithiasis. In late stages, urate deposits in the kidney may lead to chronic pyelonephritis, arteriolar sclerosis, hypertension, and renal failure.
Atlas Links	UCV1 **PM-BC-094** UCV2 IM2-052A, IM2-052B

GOUT